Sun Tzu's The Art of War Made Easy

Sun Tzu's The Art of War Made Easy

200 insightful questions and answers on
Sun Tzu's *The Art of War*

Titu Doley

First published in the United States of America in 2018

ISBN : 978-0-9996821-2-8

This book is dedicated to Sun Tzu, who authored *The Art of War* more than 2000 years ago and Lionel Giles, based on whose English translation of *The Art of War* this book has been written.

CONTENTS

PREFACE

Sun Tzu's *The Art of War* is considered one of the best books on ancient wisdom on military leadership, strategies and tactics. Though written more than 2000 years ago, it is still relevant and the principles of warfare in it can be applied to contemporary conflicts in the world.

There are numerous sources for learning Sun Tzu's philosophy of warfare in *The Art of War*. This book, *Sun Tzu's The Art of War Made Easy*, has been written to make the process of acquiring the knowledge in *The Art of War* easy. Unlike other books on the subject, this book is presented in an easy to read and grasp question and answer format.

The idea of writing this book came to me in 2017 when I was writing the Introduction section of my book titled *Mastering the Art of War by Sun Tzu* (ISBN: 9780999682111). In the Introduction section of that book I wrote some of the questions a reader will be able to answer by reading the book. I thought why not create a book which transforms the entire content of the *The Art of War* into easy to read and comprehend questions and answers. The introduction section of this book contains all the questions answered in this book. If the reader is seeking an answer to a specific question, he can quickly look up the questions in the introduction section and then go to the chapter having the answer.

I hope that you find this book valuable in quickly learning

Sun Tzu's philosophy and highly practical principles of war. Moreover, I believe that you will apply the profound knowledge gained on the art of war for only good causes. And, if you have any feedback on this book, please email me at *aowmadeeasy@gmail.com*

INTRODUCTION

Sun Tzu's *The Art of War* is a classic book on military strategies, tactics and leadership. The content in it is practical, time tested, and can be applied to war, business, politics and other competitive arenas. However, it is not an easy read and it takes a while to get the essence of Sun Tzu's philosophy and principles of warfare in it. To overcome these challenges, in this book the whole content of Sun Tzu's *The Art of War* has been organized into 200 insightful questions and answers. These easy to read and comprehend questions and answers make it easy to quickly gain the essence of Sun Tzu's *The Art of War*.

This book is organized into two parts -
Part I - In this part of the book the content of each chapter of Sun Tzu's *The Art of War* is presented in the form of questions and answers. The answers have been compiled from the translation of Sun Tzu's *The Art of War* into English by Lionel Giles.
Part II - This part contains *The Art of War* authored by Sun Tzu and translated into English by Lionel Giles.

The full repertoire of questions the reader will be able to

answer by reading the Part I of this book are given below. The following chapter-wise list of questions also give a bird's eye view of each chapter.

Chapter 1. Laying Plans

1. Why the art of war is a subject of inquiry which can on no account be neglected?
2. What are the five constant factors that govern the art of war and should be taken into account in one's deliberations, when seeking to determine the conditions in the field?
3. What does the Moral Law, a constant factor that governs the art of war, cause the people to be?
4. What does Heaven, one of the five constant factors that govern the art of war signify?
5. What is the Earth, one of the five constant factors that govern the art of war, composed of?
6. Which virtues the Commander, a constant factor that governs the art of war, stands for?
7. What is to be understood by Method and Discipline, a constant factor that governs the art of war?
8. What does a victorious general know but those who fail know not?
9. In your deliberations, when seeking to determine the military conditions, what can be made the basis of a comparison?
10. By means of which considerations, victory or defeat in war can be forecasted?
11. Which general be retained in command and which one be dismissed?
12. What did Sun Tzu say about heading his counsel and circumstances beyond the ordinary rules?

13. As all warfare is based on deception, what must we do in war?
14. Which military devices, leading to victory in war, must not be divulged beforehand ?
15. What do the winning and losing general do before the battle?

Chapter 2. Waging War

1. How much does it cost to raise an army of 100,000 men?
2. What happens in a prolonged warfare?
3. Who only can understand the profitable way of carrying on war?
4. What does a skillful soldier don't do while waging war?
5. How will the army have food enough for its needs while waging war?
6. Why does a wise general make a point of foraging on the enemy?
7. Why must our men be roused to anger in waging war?
8. Why must our men have their rewards?
9. How to use the conquered foe to augment one's own strength?
10. In war, what should be the great objective?
11. On whom depends, whether a nation shall be in peace or in peril?

Chapter 3. Attack by Stratagem

1. In the practical art of war, what is the best thing of all?
2. In the art of war, what is considered supreme excellence?
3. What are the four forms of generalship, from the highest form to the lowest form?

4. Why the rule is not to besiege walled cities if it can possibly be avoided?
5. In war, what does skillful leader do?
6. What is the rule in war with regard to the strength of our forces and the enemy's?
7. Who is the bulwark of the State and how the bulwark is related to the strength and weakness of the State?
8. What are the three ways in which a ruler can bring misfortune upon his army?
9. In the art of war, what are the five essentials for victory?
10. Who need not fear about the result of a hundred battles?
11. For every victory gained who will also suffer a defeat?
12. Who will succumb in every battle?

Chapter 4. Tactical Dispositions

1. What did the good fighters of the old do to defeat the enemy?
2. Who provides the opportunity of defeating the enemy?
3. Can a good fighter make certain of defeating the enemy?
4. What are defensive and offensive tactics?
5. When does a victorious strategist seek battle?
6. What does the consummate leader cultivate and strictly adhere to control success?
7. How is a victorious army compared to a routed one?

Chapter 5. Energy

1. What is the principle for control of large force and a few men?

2. How to fight with large army and small one under your command?
3. What are the only two methods of attack in battle?
4. What ensures that army may withstand the brunt of the enemy's attack and remain unshaken?
5. By what, the impact of attack of your army may be effected like a grindstone dashed against an egg?
6. How are good fighters in their onset and decision?
7. How does one who is skillful at keeping the enemy on the move maintain deceitful appearances, according to which the enemy will act?
8. Why does the clever combatant do not require too much energy from individuals?

Chapter 6. Weak Points and Strong

1. What can a clever combatant do to the enemy by holding out advantages?
2. When can you be sure of succeeding in your attack and safety of your defense?
3. When a general is considered skillful in attack and defense?
4. Why make for the enemy's weak points and also, be more rapid than enemy?
5. How can you force the enemy to engage in fight even though he be sheltered behind high rampart and deep ditch?
6. How to pit a whole force against the enemy's separate parts?
7. Why the spot, where we intend to attack, must not be made known to the enemy?
8. What brings numerical weakness to us and how do we gain numerical strength against the adversary?
9. How knowing the place and time of the coming battle help us?

10. How we may prevent the enemy from fighting, though he is stronger in numbers?
11. Why the tactical dispositions must be concealed to attain victory?
12. What does the multitude cannot comprehend and see about victory?
13. Should the tactics which gained you victory be repeated?
14. How military tactics are like water in its natural course?
15. Who may be called a heaven-born captain?

Chapter 7. Maneuvering

1. What must a general do before tactical maneuvering, than which there is nothing more difficult?
2. What shows the knowledge of the artifice of deviation?
3. What is the difference between maneuvering with an army and an undisciplined multitude?
4. In order to snatch an advantage, why detach a flying column?
5. What should we know before entering into alliances with neighbors?
6. We are not fit to lead an army on the march unless familiar with what?
7. Why practice dissimulation in war?
8. On what basis, whether to concentrate or to divide your troops must be decided?
9. How should your rapidity and compactness be?
10. How should raiding, plundering and immovability be?
11. How should your plans and movement be?

12. What should you do after plunder a countryside or capturing new territory?
13. What should you do before making a move?
14. Why learning the artifice of deviation is important in the art of maneuvering?
15. Why use of gongs & drums and banners & flags on the field of battle been instituted?
16. How an army be formed so that it is impossible either for the brave to advance alone or the cowardly to retreat?
17. Why use signal-fires and drums in night-fighting, and flags and banners in fighting by day?
18. Can a whole army may be robbed of its spirit and a commander-in-chief may be robbed of his presence of mind?
19. When does a clever general avoid an army and when does he attack it?
20. What is the art of retaining self-possession in warfare?
21. In war, what is the art of husbanding one's strength?
22. What is the art of studying circumstances in warfare?
23. What not to do when enemy is uphill or when he comes downhill?
24. What not to do to an enemy who simulates flight and soldiers whose temper is keen?
25. In the art of warfare, what not to do when you surround a desperate foe?

Chapter 8. Variation in Tactics

1. In war, from where the general receives his commands?

2. In the wise leader's plans, why the considerations of advantage and disadvantage of tactics be blended together?
3. In the art of war, how to handle the hostile chiefs?
4. What does the art of war teach us to rely on?
5. What are the besetting sins of a general, ruinous to the conduct of war?
6. When an army is overthrown and its leader slain, the cause will surely be found among which dangerous faults of a general?

Chapter 9. The Army on the March

1. What should the army on the march do in mountain warfare?
2. What should the army on the march do in river warfare?
3. How to do operations in salt-marshes?
4. While campaigning in flat country, where to take up position and why?
5. What are the four useful branches of military knowledge which enabled the Yellow Emperor to vanquish several sovereigns?
6. Where should the army on the march encamp if you are careful of your men and want to utilize the natural advantages of the ground?
7. What should we do in a country in which there are precipitous cliffs with torrents running between, deep natural hollows, confined places, tangled thickets, quagmires and crevasses?
8. What are the places where men in ambush or insidious spies are likely to be lurking?

9. When the enemy is close at hand and remains quiet, what is he relying on?
10. When the enemy keeps aloof and tries to provoke a battle, what is he anxious for?
11. What kind of place of encampment of enemy means he is tendering a bait?
12. What does movement amongst the trees of a forest show about the enemy?
13. What is rising of birds in their flight sign of?
14. What indicates that a sudden attack is coming?
15. What does dust rising in a high column, low dust spread over a wide area, dust branching out in different directions and few clouds of dust moving to and fro signify ?
16. What are signs that the enemy is about to advance or will retreat?
17. What is the sign that the enemy is forming for battle?
18. What do peace proposals unaccompanied by a signed covenant indicate?
19. What does it mean when there is much running about and the soldiers fall out of rank?
20. What is it when some soldiers are seen advancing and some retreating?
21. What indicates that the army is suffering from thirst?
22. What show that the enemy soldiers are exhausted?
23. If birds gather on any spot, what does it indicate?
24. If there is disturbance in the camp, what does it tell about the general's authority?
25. What indicates that sedition is afoot?
26. If the officers are angry, what does it mean?
27. What does sight of men in the rank and file whispering together in small knot or speaking in

subdued tone point to?

28. What shows a supreme lack of intelligence?
29. When envoys are sent with compliments in their mouths, what is it sign of?
30. If our troops are neither more in number nor amply sufficient than the enemy, what can we do?
31. Who is sure to be captured by opponents?
32. How soldiers must be treated and kept under control?
33. How to make an army well-disciplined?
34. When will gain be mutual for a general and his men?

Chapter 10. Terrain

1. In the art of war, what are the six kinds of terrain?
2. What is accessible ground and how to fight with advantage on accessible ground?
3. What is entangling ground and what happens if you fail to defeat enemy on entangling ground?
4. What is temporizing ground and how to deliver our attack with advantage on temporizing ground?
5. What should you do if you or enemy occupy narrow passes first?
6. What should you do if you or the adversary occupy the precipitous heights beforehand ?
7. When is it not easy to provoke a battle, and fighting will be to your disadvantage?
8. An army is exposed to which six calamities, not arising from natural causes, but from faults for which the general is responsible?
9. What are the six ways of courting defeat, which must be carefully noted by the general who has attained a responsible post?
10. What constitutes the test of a great general?

11. If fighting is sure to result in victory or defeat, then what must you do?
12. What kind of general is jewel of the kingdom?
13. How to have your soldiers follow you into the deepest valleys and stand by you even unto death?
14. What makes your soldiers useless for any practical purpose?
15. When have we gone only halfway towards victory?
16. How to make your victory not stand in doubt and make the victory complete?

Chapter 11. The Nine Situations

1. What are the nine varieties of ground that art of war recognize?
2. What is dispersive ground and what should a chieftain do on it?
3. What is facile ground and what should and shouldn't be done on it?
4. What is contentious ground and what should be done on it?
5. What is open ground and what should be done on it?
6. What is the ground of intersecting highways and what should be done on it?
7. What is serious ground and what should be done on it?
8. What is difficult ground and what should be done on it?
9. What is true of hemmed-in ground?
10. What is desperate ground and what must be done on it?
11. What did the skillful leaders of old know?
12. What did the skillful leaders of old do?

13. How to cope with a great host of the enemy in orderly array and on the point of marching to the attack?
14. What is the essence of war?
15. What are the principles to be observed by an invading force, so that without waiting to be marshaled, the soldiers will be constantly on the qui vive; without waiting to be asked, they will do your will; without restrictions, they will be faithful; without giving orders, they can be trusted ?
16. What to prohibit and do away with, to fear no calamity until death itself comes?
17. What is the principle on which to manage an army?
18. How does the skillful general conduct in the art of war?
19. At the critical moment, how does the leader of an army act?
20. What may be termed the business of the general?
21. What are the things that must most certainly be studied in the art of war?
22. When invading, what brings cohesion or dispersion?
23. When do soldiers offer an obstinate resistance, fight hard and obey promptly?
24. Until when, we can not enter into alliances with neighboring princes?
25. How does a warlike prince act in war?
26. What should and should not be known to your soldiers?
27. When is your army capable of striking a blow for victory?
28. How to forestall your opponent and time his arrival on the ground?

29. How is success gained in warfare?

Chapter 12. The Attack by Fire

1. What are the five ways of attacking with fire?
2. In order to attack with fire, what we must have available?
3. What are the special days and the proper season for making attacks with fire?
4. In attacking with fire, what are the five possible developments one should be prepared to meet?
5. Which lasts longer, a wind that rises in the daytime or a night breeze?
6. In every army, what must be known and done to attack with fire?
7. Who shows intelligence, those who use fire or water as an aid to attack?
8. In the art of war, why lay the plans well ahead and cultivate the resources?
9. In the art of war, when to make a forward move or stay where you are?
10. How to keep a country at peace and an army intact?

Chapter 13. The Use of Spies

1. How does one, who is no leader of men, no help to his sovereign and no master of victory, act?
2. What enables the wise sovereign and the good general to strike and conquer, and achieve things beyond the reach of ordinary men?
3. Why use spies in the art of war?
4. What are the five classes of spies?
5. What is the sovereign's most precious faculty?
6. How to manage and use spies for every kind of business in the art of war?

7. What to do if a spy divulges a secret piece of news before the time is ripe?
8. When do we use spies in the art of war?
9. How to make the enemy's spies available to our service?
10. Why is it essential that the converted spy be treated with the utmost liberality?
11. What do enlightened ruler and wise general use to achieve great results?
12. Why are the spies most important element in the art of war?

PART I

In this part there are questions and answers on each chapter of *The Art of War by Sun Tzu*.

1
LAYING PLANS

Q1. Why the art of war is a subject of inquiry which can on no account be neglected?

A1. Sun Tzu said: The art of war is of vital importance to the State as it is a matter of life and death, a road either to safety or to ruin. Hence it is a subject of inquiry which can on no account be neglected.

Q2. What are the five constant factors that govern the art of war and should be taken into account in one's deliberations, when seeking to determine the conditions in the field?

A2. Sun Tzu said: The five constant factors that govern the art of war and to be taken into account in one's deliberations, when seeking to determine the conditions in the field, are (1) The Moral Law; (2) Heaven; (3) Earth; (4) The Commander; (5) Method and Discipline.

Q3. What does the Moral Law, a constant factor that governs the art of war, cause the people to be?

A3. Sun Tzu said: The Moral Law causes the people to be in complete accord with their ruler, so that they will follow him regardless of their lives, undismayed by any danger.

Q4. What does Heaven, one of the five constant factors that govern the art of war signify?
A4. Sun Tzu said: Heaven signifies day and night, cold and heat, times and seasons.

Q5. What is the Earth, one of the five constant factors that govern the art of war, composed of?
A5. Sun Tzu said: Earth, one of the five constant factors that govern the art of war, comprises distances, great and small; danger and security; open ground and narrow passes; the chances of life and death.

Q6. Which virtues the Commander, a constant factor that governs the art of war, stands for?
A6. Sun Tzu said: The Commander stands for the virtues of wisdom, sincerity, benevolence, courage and strictness.

Q7. What is to be understood by Method and Discipline, a constant factor that governs the art of war?
A7. Sun Tzu said: By Method and Discipline are to be understood the marshaling of the army in its proper subdivisions, the graduations of rank among the officers, the maintenance of roads by which supplies may reach the army, and the control of military expenditure.

Q8. What does a victorious general know but those who fail know not?

A8. Sun Tzu said: The general who knows the five constant factors, (1) The Moral Law; (2) Heaven; (3) Earth; (4) The Commander; (5) Method and Discipline, will be victorious; he who knows them not will fail.

Q9. In your deliberations, when seeking to determine the military conditions, what can be made the basis of a comparison?

A9. Sun Tzu said: In your deliberations, when seeking to determine the military conditions, let these be made the basis of a comparison, in this wise:

(i) Which of the two sovereigns is imbued with the Moral law?

(ii) Which of the two generals has most ability?

(iii) With whom lie the advantages derived from Heaven and Earth?

(iv) On which side is discipline most rigorously enforced?

(v) Which army is stronger?

(vi) On which side are officers and men more highly trained?

(vii) In which army is there the greater constancy both in reward and punishment?

Q10. By means of which considerations, victory or defeat in war can be forecasted?

A10. Sun Tzu said: By means of these seven considerations I can forecast victory or defeat.

(i) Which of the two sovereigns is imbued with the Moral law?

(ii) Which of the two generals has most ability?

(iii) With whom lie the advantages derived from Heaven and Earth?

(iv) On which side is discipline most rigorously enforced?

(v) Which army is stronger?
(vi) On which side are officers and men more highly trained?
(vii) In which army is there the greater constancy both in reward and punishment?

Q11. Which general be retained in command and which one be dismissed?
A11. Sun Tzu said: The general that hearkens to my counsel and acts upon it, will conquer: let such a one be retained in command. The general that hearkens not to my counsel nor acts upon it, will suffer defeat: let such a one be dismissed.

Q12. What did Sun Tzu say about heading his counsel and circumstances beyond the ordinary rules?
A12. Sun Tzu said: While heading the profit of my counsel, avail yourself of any helpful circumstances over and beyond the ordinary rules. According as circumstances are favorable, one should modify one's plans.

Q13. As all warfare is based on deception, what must we do in war?
A13. Sun Tzu said: All warfare is based on deception. Hence,
(i) when able to attack, we must seem unable.
(ii) when using our forces, we must seem inactive.
(iii) when we are near, we must make the enemy believe we are far away.
(iv) when far away, we must make the enemy believe we are near.

Q14. Which military devices, leading to victory in war, must not be divulged beforehand ?

A14. Sun Tzu said: These military devices, leading to victory, must not be divulged beforehand –

(i) Hold out baits to entice the enemy. Feign disorder, and crush him.

(ii) If he is secure at all points, be prepared for him. If he is in superior strength, evade him.

(iii) If your opponent is of choleric temper, seek to irritate him. Pretend to be weak, that he may grow arrogant.

(iv) If he is taking his ease, give him no rest. If his forces are united, separate them.

(v) Attack him where he is unprepared, appear where you are not expected.

Q15. What do the winning and losing general do before the battle?

A15: Sun Tzu said: The general who wins a battle makes many calculations in his temple before the battle is fought. The general who loses a battle makes but few calculations beforehand. Thus, doing many calculations lead to victory, and few calculations to defeat: how much more no calculation at all! It is by attention to this point that I can foresee who is likely to win or lose.

2
WAGING WAR

Q1. How much does it cost to raise an army of 100,000 men?

A1. Sun Tzu said: In the operations of war, where there are in the field a thousand swift chariots, as many heavy chariots, and a hundred thousand mail-clad soldiers, with provisions enough to carry them a thousand li, the expenditure at home and at the front, including entertainment of guests, small items such as glue and paint, and sums spent on chariots and armor, will reach the total of a thousand ounces of silver per day. Such is the cost of raising an army of 100,000 men.

Q2. What happens in a prolonged warfare?

A2. Sun Tzu said: When you engage in actual fighting, if victory is long in coming, then men's weapons will grow dull and their ardor will be damped. If you lay siege to a town, you will exhaust your strength.

Again, if the campaign is protracted, the resources of the State will not be equal to the strain. Now, when your weapons are dulled, your ardor damped, your strength exhausted and your treasure spent, other chieftains will spring up to take advantage of your extremity. Then no man, however wise, will be able to avert the consequences that must ensue. Thus, though we have heard of stupid haste in war, cleverness has never been seen associated with long delays. There is no instance of a country having benefited from prolonged warfare.

Q3. Who only can understand the profitable way of carrying on war?
A3. Sun Tzu said: It is only one who is thoroughly acquainted with the evils of war that can thoroughly understand the profitable way of carrying it on.

Q4. What does a skillful soldier don't do while waging war?
A4. Sun Tzu said: The skillful soldier does not raise a second levy, neither are his supply-wagons loaded more than twice.

Q5. How will the army have food enough for its needs while waging war?
A5. Sun Tzu said: Bring war material with you from home, but forage on the enemy. Thus the army will have food enough for its needs.

Q6. Why does a wise general make a point of foraging on the enemy?
A6. Sun Tzu said: Poverty of the State exchequer causes an army to be maintained by contributions from a distance. Contributing to maintain an army at a distance causes the people to be impoverished.
On the other hand, the proximity of an army causes prices to go up; and high prices cause the people's substance to be drained away. When their substance is drained away, the peasantry will be afflicted by heavy exactions.
With this loss of substance and exhaustion of strength, the homes of the people will be stripped bare, and three-tenths of

their income will be dissipated; while government expenses for broken chariots, worn-out horses, breast-plates and helmets, bows and arrows, spears and shields, protective mantles, draught-oxen and heavy wagons, will amount to four-tenths of its total revenue.

Hence a wise general makes a point of foraging on the enemy. One cartload of the enemy's provisions is equivalent to twenty of one's own, and likewise a single picul of his provender is equivalent to twenty from one's own store.

Q7. Why must our men be roused to anger in waging war?

A7. Sun Tzu said: In order to kill the enemy, our men must be roused to anger.

Q8. Why must our men have their rewards?

A8. Sun Tzu said: That there may be advantage from defeating the enemy, our men must have their rewards. Therefore in chariot fighting, when ten or more chariots have been taken, those should be rewarded who took the first.

Q9. How to use the conquered foe to augment one's own strength?

A9. Sun Tzu said: In chariot fighting, when ten or more chariots have been taken, those should be rewarded who took the first. Our own flags should be substituted for those of the enemy, and the chariots mingled and used in conjunction with ours. The captured soldiers should be kindly treated and kept. This is called, using the conquered foe to augment one's own strength.

Q10. In war, what should be the great objective?

A10. Sun Tzu said: In war, let your great object be victory, not lengthy campaigns.

Q11. On whom depends, whether a nation shall be in peace or in peril?

A11. Sun Tzu said: The leader of armies is the arbiter of the people's fate, the man on whom it depends whether the nation shall be in peace or in peril.

3
ATTACK BY STRATAGEM

Q1. In the practical art of war, what is the best thing of all?
A1. Sun Tzu said: In the practical art of war, the best thing of all is to take the enemy's country whole and intact; to shatter and destroy it is not so good. So, too, it is better to recapture an army entire than to destroy it, to capture a regiment, a detachment or a company entire than to destroy them.

Q2. In the art of war, what is considered supreme excellence?
A2. Sun Tzu said: To fight and conquer in all your battles is not supreme excellence; supreme excellence consists in breaking the enemy's resistance without fighting.

Q3. What are the four forms of generalship, from the highest form to the lowest form?
A3. Sun Tzu said: The highest form of generalship is to balk the enemy's plans; the next best is to prevent the junction of the enemy's forces; the next in order is to attack the enemy's army in the field; and the worst policy of all is to besiege walled cities.

Q4. Why the rule is not to besiege walled cities if it can

possibly be avoided?
A4. Sun Tzu said: The rule is, not to besiege walled cities if it can possibly be avoided. The preparation of mantlets, movable shelters, and various implements of war, will take up three whole months. The piling up of mounds over against the walls will take three months. The general, unable to control his irritation, will launch his men to the assault like swarming ants, with the result that one-third of his men are slain, while the town still remains untaken. Such are the disastrous effects of a siege.

Q5. In war, what does skillful leader do?
A5. Sun Tzu said: The skillful leader subdues the enemy's troops without any fighting. He captures enemy cities without laying siege to them. He overthrows enemy kingdom without lengthy operations in the field.

Q6. What is the rule in war with regard to the strength of our forces and the enemy's?
A6. Sun Tzu said: It is the rule in war, if our forces are ten to the enemy's one, to surround him; if five to one, to attack him; if twice as numerous, to divide our army into two. If equally matched, we can offer battle; if slightly inferior in numbers, we can avoid the enemy; if quite unequal in every way, we can flee from him.

Q7. Who is the bulwark of the State and how the bulwark is related to the strength and weakness of the State?
A7. Sun Tzu said: The general is the bulwark of the State; if the bulwark is complete at all points; the State will be strong; if the bulwark is defective, the State will be weak.

Q8. What are the three ways in which a ruler can bring misfortune upon his army?
A8. Sun Tzu said: There are three ways in which a ruler can

bring misfortune upon his army:-

(i) By commanding the army to advance or to retreat, being ignorant of the fact that it cannot obey. This is called hobbling the army.

(ii) By attempting to govern an army in the same way as he administers a kingdom, being ignorant of the conditions which obtain in an army. This causes restlessness in the soldier's minds.

(iii) By employing the officers of his army without discrimination, through ignorance of the military principle of adaptation to circumstances. This shakes the confidence of the soldiers.

Q9. In the art of war, what are the five essentials for victory?

A9. Sun Tzu said: There are five essentials for victory:

(i) He will win who knows when to fight and when not to fight.

(ii) He will win who knows how to handle both superior and inferior forces. (iii) He will win whose army is animated by the same spirit throughout all its ranks.

(iv) He will win who, prepared himself, waits to take the enemy unprepared.

(v) He will win who has military capacity and is not interfered with by the sovereign.

Q10. Who need not fear about the result of a hundred battles?

A10. Sun Tzu said: If you know the enemy and know yourself, you need not fear the result of a hundred battles.

Q11. For every victory gained who will also suffer a defeat?

A11. Sun Tzu said: If you know yourself but not the enemy, for every victory gained you will also suffer a defeat.

Q12. Who will succumb in every battle?

A12. Sun Tzu said: If you know neither the enemy nor yourself, you will succumb in every battle.

4
TACTICAL DISPOSITIONS

Q1. What did the good fighters of the old do to defeat the enemy?
A1. Sun Tzu said: The good fighters of old first put themselves beyond the possibility of defeat, and then waited for an opportunity of defeating the enemy.

Q2. Who provides the opportunity of defeating the enemy?
A2. Sun Tzu said: To secure ourselves against defeat lies in our own hands, but the opportunity of defeating the enemy is provided by the enemy himself.

Q3. Can a good fighter make certain of defeating the enemy?
A3. Sun Tzu said: The good fighter is able to secure himself against defeat, but cannot make certain of defeating the enemy. Hence the saying: One may know how to conquer without being able to do it.

Q4. What are defensive and offensive tactics?

A4. Sun Tzu said: Security against defeat implies defensive tactics; ability to defeat the enemy means taking the offensive. Standing on the defensive indicates insufficient strength; attacking, a superabundance of strength. The general who is skilled in defense hides in the most secret recesses of the earth; he who is skilled in attack flashes forth from the topmost heights of heaven. Thus on the one hand we have ability to protect ourselves; on the other, a victory that is complete.

Q5. When does a victorious strategist seek battle?

A5. Sun Tzu said: To see victory only when it is within the ken of the common herd is not the acme of excellence. Neither is it the acme of excellence if you fight and conquer and the whole Empire says, "Well done!". To lift an autumn hair is no sign of great strength; to see the sun and moon is no sign of sharp sight; to hear the noise of thunder is no sign of a quick ear. What the ancients called a clever fighter is one who not only wins, but excels in winning with ease. Hence his victories bring him neither reputation for wisdom nor credit for courage. He wins his battles by making no mistakes. Making no mistakes is what establishes the certainty of victory, for it means conquering an enemy that is already defeated. Hence the skillful fighter puts himself into a position which makes defeat impossible, and does not miss the moment for defeating the enemy. Thus it is that in war the victorious strategist only seeks battle after the victory has been won, whereas he who is destined to defeat first fights and afterwards looks for victory.

Q6. What does the consummate leader cultivate and strictly adhere to control success?

A6. Sun Tzu said: The consummate leader cultivates the moral law, and strictly adheres to method and discipline; thus it is in his power to control success. In respect of military method, we have, firstly, Measurement; secondly, Estimation of quantity; thirdly, Calculation; fourthly, Balancing of chances; fifthly, Victory. Measurement owes its existence to Earth; Estimation of quantity to Measurement; Calculation to Estimation of quantity; Balancing of chances to Calculation; and Victory to Balancing of chances.

Q7. How is a victorious army compared to a routed one?
A7. Sun Tzu said: A victorious army opposed to a routed one, is as a pound's weight placed in the scale against a single grain.

5
ENERGY

Q1. What is the principle for control of large force and a few men?

A1. Sun Tzu said: The control of a large force is the same principle as the control of a few men: it is merely a question of dividing up their numbers.

Q2. How to fight with large army and small one under your command?

A2. Sun Tzu said: Fighting with a large army under your command is nowise different from fighting with a small one: it is merely a question of instituting signs and signals.

Q3. What are the only two methods of attack in battle?

Sun Tzu said: In battle, there are not more than two methods of attack - the direct and the indirect; yet these two in combination give rise to an endless series of maneuvers. The direct and the indirect methods of attack lead on to each other in turn. It is like moving in a circle - you never come to an end.

In all fighting, the direct method may be used for joining battle, but indirect methods will be needed in order to secure victory. Indirect tactics, efficiently applied, are inexhaustible as Heaven and Earth, unending as the flow of rivers and streams; like the sun and moon, they end but to begin anew; like the four seasons, they pass away to return once more

Q4. What ensures that army may withstand the brunt of the enemy's attack and remain unshaken?

A4. Sun Tzu said: To ensure that your whole host may withstand the brunt of the enemy's attack and remain unshaken - this is effected by direct and indirect maneuvers.

Q5. By what, the impact of attack of your army may be effected like a grindstone dashed against an egg?

A5. Sun Tzu said: That the impact of your army may be like a grindstone dashed against an egg - this is effected by the science of weak points and strong.

Q6. How are good fighters in their onset and decision?

A6. Sun Tzu said: The onset of troops is like the rush of a torrent which will even roll stones along in its course. The quality of decision is like the well-timed swoop of a falcon which enables it to strike and destroy its victim. Therefore the good fighter will be terrible in his onset, and prompt in his decision.

Q7. How does one who is skillful at keeping the enemy on the move maintain deceitful appearances, according to which the enemy will act?

A7. Sun Tzu said: Amid the turmoil and tumult of battle, there may be seeming disorder and yet no real disorder at all; amid confusion and chaos, your array may be without head or tail, yet it will be proof against defeat.

Simulated disorder postulates perfect discipline, simulated fear postulates courage; simulated weakness postulates strength.

Hiding order beneath the cloak of disorder is simply a question of subdivision; concealing courage under a show of timidity presupposes a fund of latent energy; masking strength with weakness is to be effected by tactical dispositions.

Thus one who is skillful at keeping the enemy on the move maintains deceitful appearances, according to which the enemy will act. He sacrifices something that the enemy may snatch at it.

Q8. Why does the clever combatant do not require too much energy from individuals?

A8. Sun Tzu said: The clever combatant looks to the effect of combined energy, and does not require too much from individuals. Hence his ability to pick out the right men and utilize combined energy. When he utilizes combined energy, his fighting men become as it were like unto rolling logs or stones. For it is the nature of a log or stone to remain motionless on level ground, and to move when on a slope; if four-cornered, to come to a standstill, but if round-shaped, to go rolling down. Thus the energy developed by good fighting men is as the momentum of a round stone rolled down a mountain thousands of feet in height.

6
Weak Points and Strong

Q1. What can a clever combatant do to the enemy by holding out advantages?
A1. Sun Tzu said: Whoever is first in the field and awaits the coming of the enemy, will be fresh for the fight; whoever is second in the field and has to hasten to battle will arrive exhausted. Therefore the clever combatant imposes his will on the enemy, but does not allow the enemy's will to be imposed on him. By holding out advantages to him, he can cause the enemy to approach of his own accord; or, by inflicting damage, he can make it impossible for the enemy to draw near. If the enemy is taking his ease, he can harass him; if well supplied with food, he can starve him out; if quietly encamped, he can force him to move. Appear at points which the enemy must hasten to defend; march swiftly to places where you are not expected.

Q2. When can you be sure of succeeding in your attack and safety of your defense?
A2. Sun Tzu said: You can be sure of succeeding in your attacks if you only attack places which are undefended. You

can ensure the safety of your defense if you only hold positions that cannot be attacked.

Q3. When a general is considered skillful in attack and defense?

A3. Sun Tzu said: A general is skillful in attack whose opponent does not know what to defend; and he is skillful in defense whose opponent does not know what to attack.

Q4. Why make for the enemy's weak points and also, be more rapid than enemy?

A4. Sun Tzu said: You may advance and be absolutely irresistible, if you make for the enemy's weak points; you may retire and be safe from pursuit if your movements are more rapid than those of the enemy.

Q5. How can you force the enemy to engage in fight even though he be sheltered behind high rampart and deep ditch?

A5. Sun Tzu said: If we wish to fight, the enemy can be forced to an engagement even though he be sheltered behind a high rampart and a deep ditch. All we need do is attack some other place that he will be obliged to relieve.

Q6. How to pit a whole force against the enemy's separate parts?

A6. Sun Tzu said: By discovering the enemy's dispositions and remaining invisible ourselves, we can keep our forces concentrated, while the enemy's must be divided. We can form a single united body, while the enemy must split up into fractions. Hence there will be a whole pitted against separate parts of a whole, which means that we shall be many to the enemy's few.

Q7. Why the spot, where we intend to attack, must not be made known to the enemy?

A7. Sun Tzu said: The spot where we intend to fight must not be made known; for then the enemy will have to prepare against a possible attack at several different points; and his forces being thus distributed in many directions, the numbers we shall have to face at any given point will be proportionately few.

For should the enemy strengthen his van, he will weaken his rear; should he strengthen his rear, he will weaken his van; should he strengthen his left, he will weaken his right; should he strengthen his right, he will weaken his left. If he sends reinforcements everywhere, he will everywhere be weak.

Q8. What brings numerical weakness to us and how do we gain numerical strength against the adversary?

A8. Sun Tzu said: Numerical weakness comes from having to prepare against possible attacks; numerical strength, from compelling our adversary to make these preparations against us.

Q9. How knowing the place and time of the coming battle help us?

A9. Sun Tzu said: Knowing the place and the time of the coming battle, we may concentrate from the greatest distances in order to fight. But if neither time nor place be known, then the left wing will be impotent to succor the right, the right equally impotent to succor the left, the van unable to relieve the rear, or the rear to support the van. How much more so if the furthest portions of the army are anything under a hundred LI apart, and even the nearest are separated by several LI!

Q10. How we may prevent the enemy from fighting, though he is stronger in numbers?

A10. Sun Tzu said: Though the enemy be stronger in numbers, we may prevent him from fighting.

(i) Scheme so as to discover his plans and the likelihood of their success.

(ii) Rouse him, and learn the principle of his activity or inactivity.

(iii) Force him to reveal himself, so as to find out his vulnerable spots.

(iv) Carefully compare the opposing army with your own, so that you may know where strength is superabundant and where it is deficient.

Q11. Why the tactical dispositions must be concealed to attain victory?

A11. Sun Tzu said: In making tactical dispositions, the highest pitch you can attain is to conceal them; conceal your dispositions, and you will be safe from the prying of the subtlest spies, from the machinations of the wisest brains.

Q12. What does the multitude cannot comprehend and see about victory?

A12. Sun Tzu said: How victory may be produced for them out of the enemy's own tactics - that is what the multitude cannot comprehend. All men can see the tactics whereby I conquer, but what none can see is the strategy out of which victory is evolved.

Q13. Should the tactics which gained you victory be repeated?

A13. Sun Tzu said: Do not repeat the tactics which have gained you one victory, but let your methods be regulated by the infinite variety of circumstances.

Q14. How military tactics are like water in its natural course?

A14. Sun Tzu said: Military tactics are like unto water; for water in its natural course runs away from high places and hastens downwards. So in war, the way is to avoid what is strong and to strike at what is weak. Water shapes its course according to the nature of the ground over which it flows; the soldier works out his victory in relation to the foe whom he is facing. Therefore, just as water retains no constant shape, so in warfare there are no constant conditions.

Q15. Who may be called a heaven-born captain?

A15. Sun Tzu said: He who can modify his tactics in relation to his opponent and thereby succeed in winning, may be called a heaven-born captain.

7
MANEUVERING

Q1. What must a general do before tactical maneuvering, than which there is nothing more difficult?

A1. Sun Tzu said: In war, the general having collected an army and concentrated his forces, he must blend and harmonize the different elements thereof before pitching his camp. After that, comes tactical maneuvering, than which there is nothing more difficult. The difficulty of tactical maneuvering consists in turning the devious into the direct, and misfortune into gain.

Q2. What shows the knowledge of the artifice of deviation?

A2. Sun Tzu said: To take a long and circuitous route, after enticing the enemy out of the way, and though starting after him, to contrive to reach the goal before him, shows knowledge of the artifice of deviation.

Q3. What is the difference between maneuvering with an army and an undisciplined multitude?

A3. Sun Tzu said: Maneuvering with an army is advantageous; with an undisciplined multitude, most dangerous.

Q4. In order to snatch an advantage, why detach a flying column?

A4. Sun Tzu said: If you set a fully equipped army in march in order to snatch an advantage, the chances are that you will be too late. On the other hand, to detach a flying column for the purpose involves the sacrifice of its baggage and stores. Thus, if you order your men to roll up their buff-coats, and make forced marches without halting day or night, covering double the usual distance at a stretch, doing a hundred LI in order to wrest an advantage, the leaders of all your three divisions will fall into the hands of the enemy. The stronger men will be in front, the jaded ones will fall behind, and on this plan only one-tenth of your army will reach its destination. If you march fifty LI in order to outmaneuver the enemy, you will lose the leader of your first division, and only half your force will reach the goal. If you march thirty LI with the same object, two-thirds of your army will arrive. We may take it then that an army without its baggage-train is lost; without provisions it is lost; without bases of supply it is lost.

Q5. What should we know before entering into alliances with neighbors?

A5. Sun Tzu said: We cannot enter into alliances until we are acquainted with the designs of our neighbors.

Q6. We are not fit to lead an army on the march unless familiar with what?

A6. Sun Tzu said: We are not fit to lead an army on the march unless we are familiar with the face of the country - its mountains and forests, its pitfalls and precipices, its marshes and swamps. We shall be unable to turn natural advantage to account unless we make use of local guides.

Q7. Why practice dissimulation in war?
A7. Sun Tzu said: In war, practice dissimulation, and you will succeed.

Q8. On what basis, whether to concentrate or to divide your troops must be decided?
A8. Sun Tzu said: Whether to concentrate or to divide your troops, must be decided by circumstances.

Q9. How should your rapidity and compactness be?
A9. Sun Tzu said: Let your rapidity be that of the wind, your compactness that of the forest.

Q10. How should raiding, plundering and immovability be?
A10. Sun Tzu said: In raiding and plundering be like fire, in immovability like a mountain.

Q11. How should your plans and movement be?
A11. Sun Tzu said: Let your plans be dark and impenetrable as night, and when you move, fall like a thunderbolt.

Q12. What should you do after plunder a countryside or capturing new territory?
A12. Sun Tzu said: When you plunder a countryside, let the spoil be divided amongst your men; when you capture new

territory, cut it up into allotments for the benefit of the soldiery.

Q13. What should you do before making a move?

A13. Sun Tzu said: Ponder and deliberate before you make a move.

Q14. Why learning the artifice of deviation is important in the art of maneuvering?

A14. Sun Tzu said: He will conquer who has learnt the artifice of deviation. Such is the art of maneuvering.

Q15. Why use of gongs & drums and banners & flags on the field of battle been instituted?

A15. Sun Tzu said: The Book of Army Management says: On the field of battle, the spoken word does not carry far enough: hence the institution of gongs and drums. Nor can ordinary objects be seen clearly enough: hence the institution of banners and flags.

Q16. How an army be formed so that it is impossible either for the brave to advance alone or the cowardly to retreat?

A16. Sun Tzu said: The host forming a single united body, it is impossible either for the brave to advance alone, or for the cowardly to retreat alone. This is the art of handling large masses of men.

Q17. Why use signal-fires and drums in night-fighting, and flags and banners in fighting by day?

A17. Sun Tzu said: In night-fighting make much use of signal-fires and drums, and in fighting by day, of flags and banners, as a means of influencing the ears and eyes of your army.

Q18. Can a whole army may be robbed of its spirit and a commander-in-chief may be robbed of his presence of mind?

A18. Sun Tzu said: A whole army may be robbed of its spirit; a commander-in-chief may be robbed of his presence of mind.

Q19. When does a clever general avoid an army and when does he attack it?

A19. Sun Tzu said: A soldier's spirit is keenest in the morning; by noonday it has begun to flag; and in the evening, his mind is bent only on returning to camp. A clever general, therefore, avoids an army when its spirit is keen, but attacks it when it is sluggish and inclined to return. This is the art of studying moods.

Q20. What is the art of retaining self-possession in warfare?

A20. Sun Tzu said: Disciplined and calm, to await the appearance of disorder and hubbub amongst the enemy:- this is the art of retaining self-possession.

Q21. In war, what is the art of husbanding one's strength?

A21. Sun Tzu said: To be near the goal while the enemy is still far from it, to wait at ease while the enemy is toiling and struggling, to be well-fed while the enemy is famished :- this is the art of husbanding one's strength.

Q22. What is the art of studying circumstances in warfare?

A22. Sun Tzu said: To refrain from intercepting an enemy whose banners are in perfect order, to refrain from attacking an army drawn up in calm and confident array:- this is the art of studying circumstances.

Q23. What not to do when enemy is uphill or when he comes downhill?

A23. Sun Tzu said: It is a military axiom not to advance uphill against the enemy, nor to oppose him when he comes downhill.

Q24. What not to do to an enemy who simulates flight and soldiers whose temper is keen?

A24. Sun Tzu said: Do not pursue an enemy who simulates flight; do not attack soldiers whose temper is keen.

Q25. In the art of warfare, what not to do when you surround a desperate foe?

A25. Sun Tzu said: When you surround an army, leave an outlet free. Do not press a desperate foe too hard. Such is the art of warfare.

8
VARIATION IN TACTICS

Q1. In war, from where the general receives his commands?
A1. Sun Tzu said: In war, the general receives his commands from the sovereign, collects his army and concentrates his forces.

Q2. In the wise leader's plans, why the considerations of advantage and disadvantage of tactics be blended together?
A2. Sun Tzu said: When in difficult country, do not encamp. In country where high roads intersect, join hands with your allies. Do not linger in dangerously isolated positions. In hemmed-in situations, you must resort to stratagem. In desperate position, you must fight.
There are roads which must not be followed, armies which must be not attacked, towns which must not be besieged, positions which must not be contested, commands of the sovereign which must not be obeyed.
The general who thoroughly understands the advantages that accompany variation of tactics knows how to handle his troops. The general who does not understand these, may be

well acquainted with the configuration of the country, yet he will not be able to turn his knowledge to practical account. So, the student of war who is unversed in the art of war of varying his plans will fail to make the best use of his men. Hence in the wise leader's plans, considerations of advantage and of disadvantage will be blended together. If our expectation of advantage be tempered in this way, we may succeed in accomplishing the essential part of our schemes.

Q3. In the art of war, how to handle the hostile chiefs?
A3. Sun Tzu said: Reduce the hostile chiefs by inflicting damage on them; and make trouble for them, and keep them constantly engaged; hold out specious allurements, and make them rush to any given point.

Q4. What does the art of war teach us to rely on?
A4. Sun Tzu said: The art of war teaches us to rely not on the likelihood of the enemy's not coming, but on our own readiness to receive him; not on the chance of his not attacking, but rather on the fact that we have made our position unassailable.

Q5. What are the besetting sins of a general, ruinous to the conduct of war?
A5. There are five dangerous faults which may affect a general:

(i) Recklessness, which leads to destruction;

(ii) cowardice, which leads to capture;

(iii) a hasty temper, which can be provoked by insults;

(iv) a delicacy of honor which is sensitive to shame;

(v) over-solicitude for his men, which exposes him to worry and trouble. These are the five besetting sins of a general, ruinous to the conduct of war.

Q6. When an army is overthrown and its leader slain, the cause will surely be found among which dangerous faults of a general?

A6. Sun Tzu said: When an army is overthrown and its leader slain, the cause will surely be found among these five dangerous faults which may affect a general:

(i) Recklessness, which leads to destruction;

(ii) cowardice, which leads to capture;

(iii) a hasty temper, which can be provoked by insults;

(iv) a delicacy of honor which is sensitive to shame;

(v) over-solicitude for his men, which exposes him to worry and trouble.

9
THE ARMY ON THE MARCH

Q1. What should the army on the march do in mountain warfare?

A1. Sun Tzu said: We come now to the question of encamping the army, and observing signs of the enemy. Pass quickly over mountains, and keep in the neighborhood of valleys. Camp in high places, facing the sun. Do not climb heights in order to fight. So much for mountain warfare.

Q2. What should the army on the march do in river warfare?

A2. Sun Tzu said: After crossing a river, you should get far away from it. When an invading force crosses a river in its onward march, do not advance to meet it in mid-stream. It will be best to let half the army get across, and then deliver your attack. If you are anxious to fight, you should not go to meet the invader near a river which he has to cross. Moor your craft higher up than the enemy, and facing the sun. Do not move up-stream to meet the enemy. So much for river warfare.

Q3. How to do operations in salt-marshes?

A3. Sun Tzu said: In crossing salt-marshes, your sole concern should be to get over them quickly, without any delay. If forced to fight in a salt-marsh, you should have water and grass near you, and get your back to a clump of trees. So much for operations in salt-marshes.

Q4. While campaigning in flat country, where to take up position and why?

A4. Sun Tzu said: In campaigning in dry, level country, take up an easily accessible position with rising ground to your right and on your rear, so that the danger may be in front, and safety lie behind. So much for campaigning in flat country.

Q5. What are the four useful branches of military knowledge which enabled the Yellow Emperor to vanquish several sovereigns?

A5. Sun Tzu said: These are the four useful branches of military knowledge which enabled the Yellow Emperor to vanquish several sovereigns -

(i) mountain warfare;

(ii) river warfare;

(iii) operations in salt-marshes;

(iv) campaigning in flat country.

Q6. Where should the army on the march encamp if you are careful of your men and want to utilize the natural advantages of the ground?

A6. Sun Tzu said: All armies prefer high ground to low and sunny places to dark. If you are careful of your men, and camp on hard ground, the army will be free from disease of every kind, and this will spell victory. When you come to a hill or a bank, occupy the sunny side, with the slope on your right rear. Thus you will at once act for the benefit of your soldiers and utilize the natural advantages of the ground.

Q7. What should we do in a country in which there are precipitous cliffs with torrents running between, deep natural hollows, confined places, tangled thickets, quagmires and crevasses?

A7. Sun Tzu said: Country in which there are precipitous cliffs with torrents running between, deep natural hollows, confined places, tangled thickets, quagmires and crevasses, should be left with all possible speed and not approached. While we keep away from such places, we should get the enemy to approach them; while we face them, we should let the enemy have them on his rear.

Q8. What are the places where men in ambush or insidious spies are likely to be lurking?

A8. Sun Tzu said: If in the neighborhood of your camp there should be any hilly country, ponds surrounded by aquatic grass, hollow basins filled with reeds, or woods with thick undergrowth, they must be carefully routed out and searched; for these are places where men in ambush or insidious spies are likely to be lurking.

Q9. When the enemy is close at hand and remains quiet, what is he relying on?

A9. Sun Tzu said: When the enemy is close at hand and remains quiet, he is relying on the natural strength of his position.

Q10. When the enemy keeps aloof and tries to provoke a

battle, what is he anxious for?
A10. Sun Tzu said: When he keeps aloof and tries to provoke a battle, he is anxious for the other side to advance.

Q11. What kind of place of encampment of enemy means he is tendering a bait?
A11. Sun Tzu said: If place of encampment of enemy is easy of access, he is tendering a bait.

Q12. What does movement amongst the trees of a forest show about the enemy?
A12. Sun Tzu said: Movement amongst the trees of a forest shows that the enemy is advancing.

Q13. What is rising of birds in their flight sign of?
A13. Sun Tzu said: The rising of birds in their flight is the sign of an ambuscade.

Q14. What indicates that a sudden attack is coming?
A14. Sun Tzu said: Startled beasts indicate that a sudden attack is coming.

Q15. What does dust rising in a high column, low dust spread over a wide area, dust branching out in different directions and few clouds of dust moving to and fro signify ?
A15. Sun Tzu said: When there is dust rising in a high column, it is the sign of chariots advancing; when the dust is low, but spread over a wide area, it betokens the approach of infantry. When it branches out in different directions, it shows that parties have been sent to collect firewood. A few

clouds of dust moving to and fro signify that the army is encamping.

Q16. What are signs that the enemy is about to advance or will retreat?
A16. Sun Tzu said: Humble words and increased preparations are signs that the enemy is about to advance. Violent language and driving forward as if to the attack are signs that he will retreat.

Q17. What is the sign that the enemy is forming for battle?
A17. Sun Tzu said: When the light chariots come out first and take up a position on the wings, it is a sign that the enemy is forming for battle.

Q18. What do peace proposals unaccompanied by a signed covenant indicate?
A18. Sun Tzu said: Peace proposals unaccompanied by a sworn covenant indicate a plot.

Q19. What does it mean when there is much running about and the soldiers fall out of rank?
A19. Sun Tzu said: When there is much running about and the soldiers fall out of rank, it means that the critical moment has come.

Q20. What is it when some soldiers are seen advancing and some retreating?
A20. Sun Tzu said: When some soldiers are seen advancing and some retreating, it is a lure.

Q21. What indicates that the army is suffering from thirst?

A21. Sun Tzu said: If the soldiers who are sent to draw water begin by drinking themselves, the army is suffering from thirst.

Q22. What show that the enemy soldiers are exhausted?

A22. Sun Tzu said: If the enemy sees an advantage to be gained and makes no effort to secure it, the soldiers are exhausted.

Q23. If birds gather on any spot, what does it indicate?

A23. Sun Tzu said: If birds gather on any spot, it is unoccupied.

Q24. If there is disturbance in the camp, what does it tell about the general's authority?

A24. Sun Tzu said: If there is disturbance in the camp, the general's authority is weak.

Q25. What indicates that sedition is afoot?

A25. Sun Tzu said: If the banners and flags are shifted about, sedition is afoot.

Q26. If the officers are angry, what does it mean?

A26. Sun Tzu said: If the officers are angry, it means that the men are weary.

Q27. What does sight of men in the rank and file whispering together in small knot or speaking in subdued tones point to?

A27. Sun Tzu said: The sight of men whispering together in small knots or speaking in subdued tones points to disaffection amongst the rank and file.

Q28. What shows a supreme lack of intelligence?
A28. Sun Tzu said: To begin by bluster, but afterwards to take fright at the enemy's numbers, shows a supreme lack of intelligence.

Q29. When envoys are sent with compliments in their mouths, what is it sign of?
A29. Sun Tzu said: When envoys are sent with compliments in their mouths, it is a sign that the enemy wishes for a truce.

Q30. If our troops are neither more in number nor amply sufficient than the enemy, what can we do?
A30. Sun Tzu said: If our troops are no more in number than the enemy, that is amply sufficient; it only means that no direct attack can be made. What we can do is simply to concentrate all our available strength, keep a close watch on the enemy, and obtain reinforcements.

Q31. Who is sure to be captured by opponents?
A31. Sun Tzu said: He who exercises no forethought but makes light of his opponents is sure to be captured by them.

Q32. How soldiers must be treated and kept under control?
A32. Sun Tzu said: If soldiers are punished before they have grown attached to you, they will not prove submissive; and, unless submissive, then will be practically useless. If, when

the soldiers have become attached to you, punishments are not enforced, they will still be unless.
Therefore soldiers must be treated in the first instance with humanity, but kept under control by means of iron discipline. This is a certain road to victory.

Q33. How to make an army well-disciplined?
A33. Sun Tzu said: If in training soldiers commands are habitually enforced, the army will be well-disciplined; if not, its discipline will be bad.

Q34. When will gain be mutual for a general and his men?
A34. Sun Tzu said: If a general shows confidence in his men but always insists on his orders being obeyed, the gain will be mutual.

10
TERRAIN

Q1. In the art of war, what are the six kinds of terrain?

A1. Sun Tzu said: We may distinguish six kinds of terrain, to wit:

(i) accessible ground;
(ii) entangling ground;
(iii) temporizing ground;
(iv) narrow passes;
(v) precipitous heights;
(vi) positions at a great distance from the enemy.

These are six principles connected with Earth. The general who has attained a responsible post must be careful to study them.

Q2. What is accessible ground and how to fight with advantage on accessible ground?

A2. Sun Tzu said: Ground which can be freely traversed by both sides is called accessible. With regard to ground of this nature, be before the enemy in occupying the raised and sunny spots, and carefully guard your line of supplies. Then you will be able to fight with advantage.

Q3. What is entangling ground and what happens if you fail to defeat enemy on entangling ground?
A3. Sun Tzu said: Ground which can be abandoned but is hard to re-occupy is called entangling. From a position of this sort, if the enemy is unprepared, you may sally forth and defeat him. But if the enemy is prepared for your coming, and you fail to defeat him, then, return being impossible, disaster will ensue.

Q4. What is temporizing ground and how to deliver our attack with advantage on temporizing ground?
A4. Sun Tzu said: When the position is such that neither side will gain by making the first move, it is called temporizing ground. In a position of this sort, even though the enemy should offer us an attractive bait, it will be advisable not to stir forth, but rather to retreat, thus enticing the enemy in his turn; then, when part of his army has come out, we may deliver our attack with advantage.

Q5. What should you do if you or enemy occupy narrow passes first?
A5. Sun Tzu said: With regard to narrow passes, if you can occupy them first, let them be strongly garrisoned and await the advent of the enemy. Should the enemy forestall you in occupying a pass, do not go after him if the pass is fully garrisoned, but only if it is weakly garrisoned.

Q6. What should you do if you or the adversary occupy the precipitous heights beforehand ?
A6. Sun Tzu said: With regard to precipitous heights, if you are beforehand with your adversary, you should occupy the raised and sunny spots, and there wait for him to come up. If

the enemy has occupied them before you, do not follow him, but retreat and try to entice him away.

Q7. When is it not easy to provoke a battle, and fighting will be to your disadvantage?

A7. Sun Tzu said: If you are situated at a great distance from the enemy, and the strength of the two armies is equal, it is not easy to provoke a battle, and fighting will be to your disadvantage.

Q8. An army is exposed to which six calamities, not arising from natural causes, but from faults for which the general is responsible?

A8. Sun Tzu said: An army is exposed to six several calamities, not arising from natural causes, but from faults for which the general is responsible. These are: (i) Flight; (ii) Insubordination; (iii) Collapse; (iv) Ruin; (v) Disorganization; (vi) Rout.

(i) **Flight**: Other conditions being equal, if one force is hurled against another ten times its size, the result will be the Flight of the former.

(ii) **Insubordination**: When the common soldiers are too strong and their officers too weak, the result is Insubordination.

(iii) **Collapse**: When the officers are too strong and the common soldiers too weak, the result is Collapse.

(iv) **Ruin**: When the higher officers are angry and insubordinate, and on meeting the enemy give battle on their own account from a feeling of resentment, before the commander-in-chief can tell whether or no he is in a position to fight, the result is Ruin.

(v) **Disorganization**: When the general is weak and without authority; when his orders are not clear and distinct; when there are no fixed duties assigned to officers and men, and the ranks are formed in a slovenly haphazard manner, the result is utter Disorganization.
(vi) **Rout**: When a general, unable to estimate the enemy's strength, allows an inferior force to engage a larger one, or hurls a weak detachment against a powerful one, and neglects to place picked soldiers in the front rank, the result must be Rout.

Q9. What are the six ways of courting defeat, which must be carefully noted by the general who has attained a responsible post?
A9. Sun Tzu said: These are six ways of courting defeat, which must be carefully noted by the general who has attained a responsible post.
(i)Flight; (ii) Insubordination; (iii) Collapse; (iv) Ruin; (v) Disorganization; (vi) Rout.

Q10. What constitutes the test of a great general?
A10. Sun Tzu said: The natural formation of the country is the soldier's best ally; but a power of estimating the adversary, of controlling the forces of victory, and of shrewdly calculating difficulties, dangers and distances, constitutes the test of a great general. He who knows these things, and in fighting puts his knowledge into practice, will win his battles. He who knows them not, nor practices them, will surely be defeated.

Q11. If fighting is sure to result in victory or defeat, then what must you do?

A11. Sun Tzu said: If fighting is sure to result in victory, then you must fight, even though the ruler forbid it; if fighting will not result in victory, then you must not fight even at the ruler's bidding.

Q12. What kind of general is jewel of the kingdom?
A12. Sun Tzu said: The general who advances without coveting fame and retreats without fearing disgrace, whose only thought is to protect his country and do good service for his sovereign, is the jewel of the kingdom.

Q13. How to have your soldiers follow you into the deepest valleys and stand by you even unto death?
A13. Sun Tzu said: Regard your soldiers as your children, and they will follow you into the deepest valleys; look upon them as your own beloved sons, and they will stand by you even unto death.

Q14. What makes your soldiers useless for any practical purpose?
A14. Sun Tzu said: If you are indulgent, but unable to make your authority felt; kind-hearted, but unable to enforce your commands; and incapable, moreover, of quelling disorder: then your soldiers must be likened to spoilt children; they are useless for any practical purpose.

Q15. When have we gone only halfway towards victory?
A15. Sun Tzu said: (i) If we know that our own men are in a condition to attack, but are unaware that the enemy is not open to attack, we have gone only halfway towards victory.

(ii) If we know that the enemy is open to attack, but are unaware that our own men are not in a condition to attack, we have gone only halfway towards victory.

(iii) If we know that the enemy is open to attack, and also know that our men are in a condition to attack, but are unaware that the nature of the ground makes fighting impracticable, we have still gone only halfway towards victory.

Q16. How to make your victory not stand in doubt and make the victory complete?

A16. Sun Tzu said: If you know the enemy and know yourself, your victory will not stand in doubt; if you know Heaven and know Earth, you may make your victory complete.

11
THE NINE SITUATIONS

Q1. What are the nine varieties of ground that art of war recognize?

A1. Sun Tzu said: The art of war recognizes nine varieties of ground: (i) Dispersive ground; (ii) facile ground; (iii) contentious ground; (iv) open ground; (v) ground of intersecting highways; (vi) serious ground; (vii) difficult ground; (viii) hemmed-in ground; (ix) desperate ground.

Q2. What is dispersive ground and what should a chieftain do on it?

A2. Sun Tzu said: When a chieftain is fighting in his own territory, it is dispersive ground. On dispersive ground, fight not. On dispersive ground, chieftain should inspire his men with unity of purpose.

Q3. What is facile ground and what should and shouldn't be done on it?

A3. Sun Tzu said: When chieftain has penetrated into hostile territory, but to no great distance, it is facile ground. On facile ground, halt not. On facile ground, chieftain should see that there is close connection between all parts of my army.

Q4. What is contentious ground and what should be done on it?

A4. Sun Tzu said: Ground, the possession of which imports great advantage to either side is contentious ground. On contentious ground, attack not. On contentious ground, chieftain should hurry up his men in rear.

Q5. What is open ground and what should be done on it?

A5. Sun Tzu said: Ground on which each side has liberty of movement is open ground. On open ground, do not try to block the enemy's way. On open ground, keep a vigilant eye on your defenses.

Q6. What is the ground of intersecting highways and what should be done on it?

A6. Sun Tzu said: Ground which forms the key to three contiguous states, so that he who occupies it first has most of the Empire at his command, is a ground of intersecting highways. When there are means of communication on all four sides, the ground is one of intersecting highways. On ground of intersecting highways, join hands with allies.

Q7. What is serious ground and what should be done on it?

A7. Sun Tzu said: When an army has penetrated into the heart of a hostile country, leaving a number of fortified cities in its rear, it is serious ground. On serious ground, gather in plunder. On serious ground, try to ensure a continuous stream of supplies.

Q8. What is difficult ground and what should be done on it?

A8. Sun Tzu said: Sun Tzu said: Mountain forests, rugged steeps, marshes and fens - all country that is hard to traverse: this is difficult ground. In difficult ground, keep steadily on the march. On difficult ground, don't encamp and keep pushing on along the road.

Q9. What is true of hemmed-in ground?

A9. Sun Tzu said: Ground which is reached through narrow gorges, and from which we can only retire by tortuous paths, so that a small number of the enemy would suffice to crush a large body of our men: this is hemmed in ground. When you have the enemy's strongholds on your rear, and narrow passes in front, it is hemmed-in ground. On hemmed-in ground, resort to stratagem. On hemmed-in ground block any way of retreat.

Q10. What is desperate ground and what must be done on it?

A10. Sun Tzu said: Ground on which we can only be saved from destruction by fighting without delay, is desperate ground. When there is no place of refuge at all, it is desperate ground. On desperate ground, you must fight. On desperate ground proclaim to your soldiers the hopelessness of saving their lives.

Q11. What did the skillful leaders of old know?

A11. Sun Tzu said: Those who were called skillful leaders of old knew how to -

(i) drive a wedge between the enemy's front and rear;

(ii) prevent co-operation between his large and small divisions;
(iii) hinder the good troops from rescuing the bad;
(iv) hinder the officers from rallying their men.

Q12. What did the skillful leaders of old do?
A12. Sun Tzu said: When the enemy's men were united, they managed to keep them in disorder. When it was to their advantage, they made a forward move; when otherwise, they stopped still.

Q13. How to cope with a great host of the enemy in orderly array and on the point of marching to the attack?
A13. Sun Tzu said: If asked how to cope with a great host of the enemy in orderly array and on the point of marching to the attack, I should say: "Begin by seizing something which your opponent holds dear; then he will be amenable to your will."

Q14. What is the essence of war?
A14. Sun Tzu said: Rapidity is the essence of war: take advantage of the enemy's unreadiness, make your way by unexpected routes, and attack unguarded spots.

Q15. What are the principles to be observed by an invading force, so that without waiting to be marshaled, the soldiers will be constantly on the qui vive; without waiting to be asked, they will do your will; without restrictions, they will be faithful; without giving orders, they can be trusted?

A15. Sun Tzu said: The following are the principles to be observed by an invading force:
(i) The further you penetrate into a country, the greater will be the solidarity of your troops, and thus the defenders will not prevail against you.
(ii) Make forays in fertile country in order to supply your army with food.
(iii) Carefully study the well-being of your men, and do not overtax them. Concentrate your energy and hoard your strength. Keep your army continually on the move, and devise unfathomable plans.
(iv) Throw your soldiers into positions whence there is no escape, and they will prefer death to flight. If they will face death, there is nothing they may not achieve. Officers and men alike will put forth their uttermost strength. Soldiers when in desperate straits lose the sense of fear. If there is no place of refuge, they will stand firm. If they are in hostile country, they will show a stubborn front. If there is no help for it, they will fight hard.

Thus, without waiting to be marshaled, the soldiers will be constantly on the qui vive; without waiting to be asked, they will do your will; without restrictions, they will be faithful; without giving orders, they can be trusted.

Q16. What to prohibit and do away with, to fear no calamity until death itself comes?
A16. Sun Tzu said: Prohibit the taking of omens, and do away with superstitious doubts. Then, until death itself comes, no calamity need be feared.

Q17. What is the principle on which to manage an army?

A17. Sun Tzu said: The principle on which to manage an army is to set up one standard of courage which all must reach.

Q18. How does the skillful general conduct in the art of war?

A18. Sun Tzu said: The skillful general conducts his army just as though he were leading a single man, willy-nilly, by the hand. It is the business of a general to be quiet and thus ensure secrecy; upright and just, and thus maintain order. He must be able to mystify his officers and men by false reports and appearances, and thus keep them in total ignorance. By altering his arrangements and changing his plans, he keeps the enemy without definite knowledge. By shifting his camp and taking circuitous routes, he prevents the enemy from anticipating his purpose.

Q19. At the critical moment, how does the leader of an army act?

A19. Sun Tzu said: At the critical moment, the leader of an army acts like one who has climbed up a height and then kicks away the ladder behind him. He carries his men deep into hostile territory before he shows his hand. He burns his boats and breaks his cooking-pots. Like a shepherd driving a flock of sheep, he drives his men this way and that, and nothing knows whither he is going.

Q20. What may be termed the business of the general?

A20. Sun Tzu said: To muster his host and bring it into danger:- this may be termed the business of the general.

Q21. What are the things that must most certainly be studied in the art of war?

A21. Sun Tzu said: The different measures suited to the nine varieties of ground; the expediency of aggressive or defensive tactics; and the fundamental laws of human nature: these are things that must most certainly be studied.

Q22. When invading, what brings cohesion or dispersion?

A22. Sun Tzu said: When invading hostile territory, the general principle is, that penetrating deeply brings cohesion; penetrating but a short way means dispersion.

Q23. When do soldiers offer an obstinate resistance, fight hard and obey promptly?

A23. Sun Tzu said: It is the soldier's disposition to offer an obstinate resistance when surrounded, to fight hard when he cannot help himself, and to obey promptly when he has fallen into danger.

Q24. Until when, we can not enter into alliances with neighboring princes?

A24. Sun Tzu said: We cannot enter into alliance with neighboring princes until we are acquainted with their designs.

Q25. How does a warlike prince act in war?

A25. Sun Tzu said: When a warlike prince attacks a powerful state, his generalship shows itself in preventing the concentration of the enemy's forces. He overawes his opponents, and their allies are prevented from joining against him. He does not strive to ally himself with all and sundry, nor does he foster the power of other states. He carries out his own secret designs, keeping his antagonists in awe. Thus he is able to capture their cities and overthrow their kingdoms.

Q26. What should and should not be known to your soldiers?

A26. Sun Tzu said: Confront your soldiers with the deed itself; never let them know your design. When the outlook is bright, bring it before their eyes; but tell them nothing when the situation is gloomy.

Q27. When is your army capable of striking a blow for victory?

A27. Sun Tzu said: Place your army in deadly peril, and it will survive; plunge it into desperate straits, and it will come off in safety. For it is precisely when a force has fallen into harm's way that is capable of striking a blow for victory.

Q28. How to forestall your opponent and time his arrival on the ground?

A28. Sun Tzu said: Forestall your opponent by seizing what he holds dear, and subtly contrive to time his arrival on the ground.

Q29. How is success gained in warfare?

A29. Sun Tzu said: Success in warfare is gained by carefully accommodating ourselves to the enemy's purpose. Walk in the path defined by rule, and accommodate yourself to the enemy until you can fight a decisive battle. At first exhibit the coyness of a maiden, until the enemy gives you an opening; afterwards emulate the rapidity of a running hare, and it will be too late for the enemy to oppose you.

12
The Attack by Fire

Q1. What are the five ways of attacking with fire?
A1. Sun Tzu said: There are five ways of attacking with fire. The first is to burn soldiers in their camp; the second is to burn stores; the third is to burn baggage trains; the fourth is to burn arsenals and magazines; the fifth is to hurl dropping fire amongst the enemy.

Q2. In order to attack with fire, what we must have available?
A2. Sun Tzu said: In order to carry out an attack, we must have means available. The material for raising fire should always be kept in readiness.

Q3. What are the special days and the proper season for making attacks with fire?
A3. Sun Tzu said: There is a proper season for making attacks with fire, and special days for starting a conflagration. The proper season is when the weather is very dry; the special days are those when the moon is in the

constellations of the Sieve, the Wall, the Wing or the Cross-bar; for these four are all days of rising wind.

Q4. In attacking with fire, what are the five possible developments one should be prepared to meet?

A4. Sun Tzu said: In attacking with fire, one should be prepared to meet five possible developments:

(i) When fire breaks out inside to enemy's camp, respond at once with an attack from without.

(ii) If there is an outbreak of fire, but the enemy's soldiers remain quiet, bide your time and do not attack.

(iii) When the force of the flames have reached its height, follow it up with an attack, if that is practicable; if not, stay where you are.

(iv) If it is possible to make an assault with fire from without, do not wait for it to break out within, but deliver your attack at a favorable moment.

(v) When you start a fire, be to windward of it. Do not attack from the leeward.

Q5. Which lasts longer, a wind that rises in the daytime or a night breeze?

A5. Sun Tzu said: A wind that rises in the daytime lasts long, but a night breeze soon falls.

Q6. In every army, what must be known and done to attack with fire?

A6. Sun Tzu said: In every army, the five possible developments connected with fire must be known, the movements of the stars calculated, and a watch kept for the proper days.

Q7. Who shows intelligence, those who use fire or water as an aid to attack?

A7. Sun Tzu said: Those who use fire as an aid to the attack show intelligence; those who use water as an aid to the attack gain an accession of strength. By means of water, an enemy may be intercepted, but not robbed of all his belongings.

Q8. In the art of war, why lay the plans well ahead and cultivate the resources?

A8. Sun Tzu said: Unhappy is the fate of one who tries to win his battles and succeed in his attacks without cultivating the spirit of enterprise; for the result is waste of time and general stagnation. Hence the saying: The enlightened ruler lays his plans well ahead; the good general cultivates his resources.

Q9. In the art of war, when to make a forward move or stay where you are?

A9. Sun Tzu said: Move not unless you see an advantage; use not your troops unless there is something to be gained; fight not unless the position is critical. No ruler should put troops into the field merely to gratify his own spleen; no general should fight a battle simply out of pique. If it is to your advantage, make a forward move; if not, stay where you are.

Q10. How to keep a country at peace and an army intact?

A10. Sun Tzu said: Anger may in time change to gladness; vexation may be succeeded by content. But a kingdom that has once been destroyed can never come again into being;

nor can the dead ever be brought back to life. Hence the enlightened ruler is heedful, and the good general full of caution. This is the way to keep a country at peace and an army intact.

13
THE USE OF SPIES

Q1. How does one, who is no leader of men, no help to his sovereign and no master of victory, act?

A1. Sun Tzu said: Raising a host of a hundred thousand men and marching them great distances entails heavy loss on the people and a drain on the resources of the State. The daily expenditure will amount to a thousand ounces of silver. There will be commotion at home and abroad, and men will drop down exhausted on the highways. As many as seven hundred thousand families will be impeded in their labor.

Hostile armies may face each other for years, striving for the victory which is decided in a single day. This being so, to remain in ignorance of the enemy's condition simply because one grudges the outlay of a hundred ounces of silver in honors and emoluments, is the height of inhumanity.

One who acts thus is no leader of men, no present help to his sovereign, no master of victory.

Q2. What enables the wise sovereign and the good general to strike and conquer, and achieve things beyond the reach of ordinary men?

A2. Sun Tzu said: What enables the wise sovereign and the good general to strike and conquer, and achieve things beyond the reach of ordinary men, is foreknowledge.

Q3. Why use spies in the art of war?
A3. Sun Tzu said: What enables the wise sovereign and the good general to strike and conquer, and achieve things beyond the reach of ordinary men, is foreknowledge. Now this foreknowledge cannot be elicited from spirits; it cannot be obtained inductively from experience, nor by any deductive calculation. Knowledge of the enemy's dispositions can only be obtained from other men. Hence the use of spies.

Q4. What are the five classes of spies?
A4. Sun Tzu said: The five classes of spies are :
(i) **Local spies**: Having local spies means employing the services of the inhabitants of a district.
(ii) **Inward spies**: Having inward spies means making use of officials of the enemy.
(iii) **Converted spies**: Having converted spies means getting hold of the enemy's spies and using them for our own purposes.
(iv) **Doomed spies**: Having doomed spies means doing certain things openly for purposes of deception, and allowing our spies to know of them and report them to the enemy.
(v) **Surviving spies**: Surviving spies are those who bring back news from the enemy's camp.

Q5. What is the sovereign's most precious faculty?
A5. Sun Tzu said: When five kinds of spies are all at work, none can discover the secret system. This is called "divine

manipulation of the threads." It is the sovereign's most precious faculty.

Q6. How to manage and use spies for every kind of business in the art of war?

A6. Sun Tzu said: None in the whole army are more intimate relations to be maintained than with spies. None should be more liberally rewarded than the spies. In no other business should greater secrecy be preserved. Spies cannot be usefully employed without a certain intuitive sagacity. They cannot be properly managed without benevolence and straightforwardness. Without subtle ingenuity of mind, one cannot make certain of the truth of their reports. Be subtle! be subtle! and use your spies for every kind of business.

Q7. What to do if a spy divulges a secret piece of news before the time is ripe?

A7. Sun Tzu said: If a secret piece of news is divulged by a spy before the time is ripe, he must be put to death together with the man to whom the secret was told.

Q8. When do we use spies in the art of war?

A8. Sun Tzu said: Whether the object be to crush an army, to storm a city, or to assassinate an individual, it is always necessary to begin by finding out the names of the attendants, the aides-de-camp, and door-keepers and sentries of the general in command. Our spies must be commissioned to ascertain these.

Q9. How to make the enemy's spies available to our service?

A9. Sun Tzu said: The enemy's spies who have come to spy on us must be sought out, tempted with bribes, led away and comfortably housed. Thus they will become converted spies and available for our service.

Q10. Why is it essential that the converted spy be treated with the utmost liberality?

A10. Sun Tzu said: It is essential that the converted spy be treated with the utmost liberality because:

(i) It is through the information brought by the converted spy that we are able to acquire and employ local and inward spies.

(ii) It is owing to the information brought by the converted spy that we can cause the doomed spy to carry false tidings to the enemy.

(iii) It is by information brought by the converted spy that the surviving spy can be used on appointed occasions.

(iv) The end and aim of spying in all its five varieties is knowledge of the enemy; and this knowledge can only be derived, in the first instance, from the converted spy.

Q11. What do enlightened ruler and wise general use to achieve great results?

A11. Sun Tzu said: It is only the enlightened ruler and the wise general who will use the highest intelligence of the army for purposes of spying and thereby they achieve great results.

Q12. Why are the spies most important element in the art of war?

A12. Sun Tzu said: Spies are a most important element in war, because on them depends an army's ability to move.

PART II

This part contains *The Art of War* authored by Sun Tzu and translated into English by Lionel Giles.

The Art of War

By Sun Tzu

Translated By Lionel Giles

1
LAYING PLANS

1. Sun Tzu said: The art of war is of vital importance to the State.

2. It is a matter of life and death, a road either to safety or to ruin. Hence it is a subject of inquiry which can on no account be neglected.

3. The art of war, then, is governed by five constant factors, to be taken into account in one's deliberations, when seeking to determine the conditions obtaining in the field.

4. These are: (1) The Moral Law; (2) Heaven; (3) Earth; (4) The Commander; (5) Method and Discipline.

5. The Moral Law causes the people to be in complete accord with their ruler, so that they will follow him regardless of their lives, undismayed by any danger.

6. Heaven signifies night and day, cold and heat, times and seasons.

7. Earth comprises distances, great and small; danger and security; open ground and narrow passes; the chances of life and death.

8. The Commander stands for the virtues of wisdom, sincerity, benevolence, courage and strictness.

9. By Method and Discipline are to be understood the marshaling of the army in its proper subdivisions, the graduations of rank among the officers, the maintenance of roads by which supplies may reach the army, and the control of military expenditure.

10. These five heads should be familiar to every general: he who knows them will be victorious; he who knows them not will fail.

11. Therefore, in your deliberations, when seeking to determine the military conditions, let them be made the basis of a comparison, in this wise:-

12. (1) Which of the two sovereigns is imbued with the Moral law? (2) Which of the two generals has most ability? (3) With whom lie the advantages derived from Heaven and Earth? (4) On which side is discipline most rigorously enforced? (5) Which army is stronger? (6) On which side are officers and men more highly trained? (7) In which army is there the greater constancy both in reward and punishment?

13. By means of these seven considerations I can forecast victory or defeat.

14. The general that hearkens to my counsel and acts upon it, will conquer: let such a one be retained in command! The general that hearkens not to my counsel nor acts upon it, will suffer defeat:- let such a one be dismissed!

15. While heading the profit of my counsel, avail yourself also of any helpful circumstances over and beyond the ordinary rules.

16. According as circumstances are favorable, one should modify one's plans.

17. All warfare is based on deception.

18. Hence, when able to attack, we must seem unable; when using our forces, we must seem inactive; when we are near, we must make the enemy believe we are far away; when far away, we must make him believe we are near.

19. Hold out baits to entice the enemy. Feign disorder, and crush him.

20. If he is secure at all points, be prepared for him. If he is in superior strength, evade him.

21. If your opponent is of choleric temper, seek to irritate him. Pretend to be weak, that he may grow arrogant.

22. If he is taking his ease, give him no rest. If his forces are united, separate them.

23. Attack him where he is unprepared, appear where you are not expected.

24. These military devices, leading to victory, must not be divulged beforehand.

25. Now the general who wins a battle makes many calculations in his temple ere the battle is fought. The general who loses a battle makes but few calculations beforehand. Thus do many calculations lead to victory, and few calculations to defeat: how much more no calculation at all! It is by attention to this point that I can foresee who is likely to win or lose.

2
WAGING WAR

1. Sun Tzu said: In the operations of war, where there are in the field a thousand swift chariots, as many heavy chariots, and a hundred thousand mail-clad soldiers, with provisions enough to carry them a thousand li, the expenditure at home and at the front, including entertainment of guests, small items such as glue and paint, and sums spent on chariots and armor, will reach the total of a thousand ounces of silver per day. Such is the cost of raising an army of 100,000 men.

2. When you engage in actual fighting, if victory is long in coming, then men's weapons will grow dull and their ardor will be damped. If you lay siege to a town, you will exhaust your strength.

3. Again, if the campaign is protracted, the resources of the State will not be equal to the strain.

4. Now, when your weapons are dulled, your ardor damped, your strength exhausted and your treasure spent, other chieftains will spring up to take advantage of your extremity.

Then no man, however wise, will be able to avert the consequences that must ensue.

5. Thus, though we have heard of stupid haste in war, cleverness has never been seen associated with long delays.

6. There is no instance of a country having benefited from prolonged warfare.

7. It is only one who is thoroughly acquainted with the evils of war that can thoroughly understand the profitable way of carrying it on.

8. The skillful soldier does not raise a second levy, neither are his supply-wagons loaded more than twice.

9. Bring war material with you from home, but forage on the enemy. Thus the army will have food enough for its needs.

10. Poverty of the State exchequer causes an army to be maintained by contributions from a distance. Contributing to maintain an army at a distance causes the people to be impoverished.

11. On the other hand, the proximity of an army causes prices to go up; and high prices cause the people's substance to be drained away.

12. When their substance is drained away, the peasantry will be afflicted by heavy exactions.

13. With this loss of substance and exhaustion of strength, the homes of the people will be stripped bare, and three-tenths of their income will be dissipated; while government expenses for broken chariots, worn-out horses, breast-plates and helmets, bows and arrows, spears and shields, protective mantles, draught-oxen and heavy wagons, will amount to four-tenths of its total revenue.

14. Hence a wise general makes a point of foraging on the enemy. One cartload of the enemy's provisions is equivalent to twenty of one's own, and likewise a single picul of his provender is equivalent to twenty from one's own store.

15. Now in order to kill the enemy, our men must be roused to anger; that there may be advantage from defeating the enemy, they must have their rewards.

16. Therefore in chariot fighting, when ten or more chariots have been taken, those should be rewarded who took the first. Our own flags should be substituted for those of the enemy, and the chariots mingled and used in conjunction with ours. The captured soldiers should be kindly treated and kept.

17. This is called, using the conquered foe to augment one's own strength.

18. In war, then, let your great object be victory, not lengthy campaigns.

19. Thus it may be known that the leader of armies is the arbiter of the people's fate, the man on whom it depends whether the nation shall be in peace or in peril.

3
ATTACK BY STRATAGEM

1. Sun Tzu said: In the practical art of war, the best thing of all is to take the enemy's country whole and intact; to shatter and destroy it is not so good. So, too, it is better to recapture an army entire than to destroy it, to capture a regiment, a detachment or a company entire than to destroy them.

2. Hence to fight and conquer in all your battles is not supreme excellence; supreme excellence consists in breaking the enemy's resistance without fighting.

3. Thus the highest form of generalship is to balk the enemy's plans; the next best is to prevent the junction of the enemy's forces; the next in order is to attack the enemy's army in the field; and the worst policy of all is to besiege walled cities.

4. The rule is, not to besiege walled cities if it can possibly be avoided. The preparation of mantlets, movable shelters, and various implements of war, will take up three whole

months; and the piling up of mounds over against the walls will take three months more.

5. The general, unable to control his irritation, will launch his men to the assault like swarming ants, with the result that one-third of his men are slain, while the town still remains untaken. Such are the disastrous effects of a siege.

6. Therefore the skillful leader subdues the enemy's troops without any fighting; he captures their cities without laying siege to them; he overthrows their kingdom without lengthy operations in the field.

7. With his forces intact he will dispute the mastery of the Empire, and thus, without losing a man, his triumph will be complete. This is the method of attacking by stratagem.

8. It is the rule in war, if our forces are ten to the enemy's one, to surround him; if five to one, to attack him; if twice as numerous, to divide our army into two.

9. If equally matched, we can offer battle; if slightly inferior in numbers, we can avoid the enemy; if quite unequal in every way, we can flee from him.

10. Hence, though an obstinate fight may be made by a small force, in the end it must be captured by the larger force.

11. Now the general is the bulwark of the State; if the bulwark is complete at all points; the State will be strong; if the bulwark is defective, the State will be weak.

12. There are three ways in which a ruler can bring misfortune upon his army:-

13. (1) By commanding the army to advance or to retreat, being ignorant of the fact that it cannot obey. This is called hobbling the army.

14. (2) By attempting to govern an army in the same way as he administers a kingdom, being ignorant of the conditions which obtain in an army. This causes restlessness in the soldier's minds.

15. (3) By employing the officers of his army without discrimination, through ignorance of the military principle of adaptation to circumstances. This shakes the confidence of the soldiers.

16. But when the army is restless and distrustful, trouble is sure to come from the other feudal princes. This is simply bringing anarchy into the army, and flinging victory away.

17. Thus we may know that there are five essentials for victory: (1) He will win who knows when to fight and when not to fight. (2) He will win who knows how to handle both superior and inferior forces. (3) He will win whose army is animated by the same spirit throughout all its ranks. (4) He will win who, prepared himself, waits to take the enemy unprepared. (5) He will win who has military capacity and is not interfered with by the sovereign.

18. Hence the saying: If you know the enemy and know yourself, you need not fear the result of a hundred battles. If

you know yourself but not the enemy, for every victory gained you will also suffer a defeat. If you know neither the enemy nor yourself, you will succumb in every battle.

4
TACTICAL DISPOSITIONS

1. Sun Tzu said: The good fighters of old first put themselves beyond the possibility of defeat, and then waited for an opportunity of defeating the enemy.

2. To secure ourselves against defeat lies in our own hands, but the opportunity of defeating the enemy is provided by the enemy himself.

3. Thus the good fighter is able to secure himself against defeat, but cannot make certain of defeating the enemy.

4. Hence the saying: One may know how to conquer without being able to do it.

5. Security against defeat implies defensive tactics; ability to defeat the enemy means taking the offensive.

6. Standing on the defensive indicates insufficient strength; attacking, a superabundance of strength.

7. The general who is skilled in defense hides in the most secret recesses of the earth; he who is skilled in attack flashes forth from the topmost heights of heaven. Thus on the one hand we have ability to protect ourselves; on the other, a victory that is complete.

8. To see victory only when it is within the ken of the common herd is not the acme of excellence.

9. Neither is it the acme of excellence if you fight and conquer and the whole Empire says, "Well done!"

10. To lift an autumn hair is no sign of great strength; to see the sun and moon is no sign of sharp sight; to hear the noise of thunder is no sign of a quick ear.

11. What the ancients called a clever fighter is one who not only wins, but excels in winning with ease.

12. Hence his victories bring him neither reputation for wisdom nor credit for courage.

13. He wins his battles by making no mistakes. Making no mistakes is what establishes the certainty of victory, for it means conquering an enemy that is already defeated.

14. Hence the skillful fighter puts himself into a position which makes defeat impossible, and does not miss the moment for defeating the enemy.

15. Thus it is that in war the victorious strategist only seeks battle after the victory has been won, whereas he who is

destined to defeat first fights and afterwards looks for victory.

16. The consummate leader cultivates the moral law, and strictly adheres to method and discipline; thus it is in his power to control success.

17. In respect of military method, we have, firstly, Measurement; secondly, Estimation of quantity; thirdly, Calculation; fourthly, Balancing of chances; fifthly, Victory.

18. Measurement owes its existence to Earth; Estimation of quantity to Measurement; Calculation to Estimation of quantity; Balancing of chances to Calculation; and Victory to Balancing of chances.

19. A victorious army opposed to a routed one, is as a pound's weight placed in the scale against a single grain.

20. The onrush of a conquering force is like the bursting of pent-up waters into a chasm a thousand fathoms deep.

5
ENERGY

1. Sun Tzu said: The control of a large force is the same principle as the control of a few men: it is merely a question of dividing up their numbers.

2. Fighting with a large army under your command is nowise different from fighting with a small one: it is merely a question of instituting signs and signals.

3. To ensure that your whole host may withstand the brunt of the enemy's attack and remain unshaken - this is effected by maneuvers direct and indirect.

4. That the impact of your army may be like a grindstone dashed against an egg - this is effected by the science of weak points and strong.

5. In all fighting, the direct method may be used for joining battle, but indirect methods will be needed in order to secure victory.

6. Indirect tactics, efficiently applied, are inexhaustible as Heaven and Earth, unending as the flow of rivers and streams; like the sun and moon, they end but to begin anew; like the four seasons, they pass away to return once more.

7. There are not more than five musical notes, yet the combinations of these five give rise to more melodies than can ever be heard.

8. There are not more than five primary colors (blue, yellow, red, white, and black), yet in combination they produce more hues than can ever been seen.

9. There are not more than five cardinal tastes (sour, acrid, salt, sweet, bitter), yet combinations of them yield more flavors than can ever be tasted.

10. In battle, there are not more than two methods of attack--the direct and the indirect; yet these two in combination give rise to an endless series of maneuvers.

11. The direct and the indirect lead on to each other in turn. It is like moving in a circle - you never come to an end. Who can exhaust the possibilities of their combination?

12. The onset of troops is like the rush of a torrent which will even roll stones along in its course.

13. The quality of decision is like the well-timed swoop of a falcon which enables it to strike and destroy its victim.

14. Therefore the good fighter will be terrible in his onset, and prompt in his decision.

15. Energy may be likened to the bending of a crossbow; decision, to the releasing of a trigger.

16. Amid the turmoil and tumult of battle, there may be seeming disorder and yet no real disorder at all; amid confusion and chaos, your array may be without head or tail, yet it will be proof against defeat.

17. Simulated disorder postulates perfect discipline, simulated fear postulates courage; simulated weakness postulates strength.

18. Hiding order beneath the cloak of disorder is simply a question of subdivision; concealing courage under a show of timidity presupposes a fund of latent energy; masking strength with weakness is to be effected by tactical dispositions.

19. Thus one who is skillful at keeping the enemy on the move maintains deceitful appearances, according to which the enemy will act. He sacrifices something, that the enemy may snatch at it.

20. By holding out baits, he keeps him on the march; then with a body of picked men he lies in wait for him.

21. The clever combatant looks to the effect of combined energy, and does not require too much from individuals.

Hence his ability to pick out the right men and utilize combined energy.

22. When he utilizes combined energy, his fighting men become as it were like unto rolling logs or stones. For it is the nature of a log or stone to remain motionless on level ground, and to move when on a slope; if four-cornered, to come to a standstill, but if round-shaped, to go rolling down.

23. Thus the energy developed by good fighting men is as the momentum of a round stone rolled down a mountain thousands of feet in height. So much on the subject of energy.

6
WEAK POINTS AND STRONG

1. Sun Tzu said: Whoever is first in the field and awaits the coming of the enemy, will be fresh for the fight; whoever is second in the field and has to hasten to battle will arrive exhausted.

2. Therefore the clever combatant imposes his will on the enemy, but does not allow the enemy's will to be imposed on him.

3. By holding out advantages to him, he can cause the enemy to approach of his own accord; or, by inflicting damage, he can make it impossible for the enemy to draw near.

4. If the enemy is taking his ease, he can harass him; if well supplied with food, he can starve him out; if quietly encamped, he can force him to move.

5. Appear at points which the enemy must hasten to defend; march swiftly to places where you are not expected.

6. An army may march great distances without distress, if it marches through country where the enemy is not.

7. You can be sure of succeeding in your attacks if you only attack places which are undefended. You can ensure the safety of your defense if you only hold positions that cannot be attacked.

8. Hence that general is skillful in attack whose opponent does not know what to defend; and he is skillful in defense whose opponent does not know what to attack.

9. O divine art of subtlety and secrecy! Through you we learn to be invisible, through you inaudible; and hence we can hold the enemy's fate in our hands.

10. You may advance and be absolutely irresistible, if you make for the enemy's weak points; you may retire and be safe from pursuit if your movements are more rapid than those of the enemy.

11. If we wish to fight, the enemy can be forced to an engagement even though he be sheltered behind a high rampart and a deep ditch. All we need do is attack some other place that he will be obliged to relieve.

12. If we do not wish to fight, we can prevent the enemy from engaging us even though the lines of our encampment be merely traced out on the ground. All we need do is to throw something odd and unaccountable in his way.

13. By discovering the enemy's dispositions and remaining invisible ourselves, we can keep our forces concentrated, while the enemy's must be divided.

14. We can form a single united body, while the enemy must split up into fractions. Hence there will be a whole pitted against separate parts of a whole, which means that we shall be many to the enemy's few.

15. And if we are able thus to attack an inferior force with a superior one, our opponents will be in dire straits.

16. The spot where we intend to fight must not be made known; for then the enemy will have to prepare against a possible attack at several different points; and his forces being thus distributed in many directions, the numbers we shall have to face at any given point will be proportionately few.

17. For should the enemy strengthen his van, he will weaken his rear; should he strengthen his rear, he will weaken his van; should he strengthen his left, he will weaken his right; should he strengthen his right, he will weaken his left. If he sends reinforcements everywhere, he will everywhere be weak.

18. Numerical weakness comes from having to prepare against possible attacks; numerical strength, from compelling our adversary to make these preparations against us.

19. Knowing the place and the time of the coming battle, we may concentrate from the greatest distances in order to fight.

20. But if neither time nor place be known, then the left wing will be impotent to succor the right, the right equally impotent to succor the left, the van unable to relieve the rear, or the rear to support the van. How much more so if the furthest portions of the army are anything under a hundred LI apart, and even the nearest are separated by several LI!

21. Though according to my estimate the soldiers of Yueh exceed our own in number, that shall advantage them nothing in the matter of victory. I say then that victory can be achieved.

22. Though the enemy be stronger in numbers, we may prevent him from fighting. Scheme so as to discover his plans and the likelihood of their success.

23. Rouse him, and learn the principle of his activity or inactivity. Force him to reveal himself, so as to find out his vulnerable spots.

24. Carefully compare the opposing army with your own, so that you may know where strength is superabundant and where it is deficient.

25. In making tactical dispositions, the highest pitch you can attain is to conceal them; conceal your dispositions, and you will be safe from the prying of the subtlest spies, from the machinations of the wisest brains.

26. How victory may be produced for them out of the enemy's own tactics--that is what the multitude cannot comprehend.

27. All men can see the tactics whereby I conquer, but what none can see is the strategy out of which victory is evolved.

28. Do not repeat the tactics which have gained you one victory, but let your methods be regulated by the infinite variety of circumstances.

29. Military tactics are like unto water; for water in its natural course runs away from high places and hastens downwards.

30. So in war, the way is to avoid what is strong and to strike at what is weak.

31. Water shapes its course according to the nature of the ground over which it flows; the soldier works out his victory in relation to the foe whom he is facing.

32. Therefore, just as water retains no constant shape, so in warfare there are no constant conditions.

33. He who can modify his tactics in relation to his opponent and thereby succeed in winning, may be called a heaven-born captain.

34. The five elements (water, fire, wood, metal, earth) are not always equally predominant; the four seasons make way

for each other in turn. There are short days and long; the moon has its periods of waning and waxing.

7
MANEUVERING

1. Sun Tzu said: In war, the general receives his commands from the sovereign.

2. Having collected an army and concentrated his forces, he must blend and harmonize the different elements thereof before pitching his camp.

3. After that, comes tactical maneuvering, than which there is nothing more difficult. The difficulty of tactical maneuvering consists in turning the devious into the direct, and misfortune into gain.

4. Thus, to take a long and circuitous route, after enticing the enemy out of the way, and though starting after him, to contrive to reach the goal before him, shows knowledge of the artifice of deviation.

5. Maneuvering with an army is advantageous; with an undisciplined multitude, most dangerous.

6. If you set a fully equipped army in march in order to snatch an advantage, the chances are that you will be too late. On the other hand, to detach a flying column for the purpose involves the sacrifice of its baggage and stores.

7. Thus, if you order your men to roll up their buff-coats, and make forced marches without halting day or night, covering double the usual distance at a stretch, doing a hundred LI in order to wrest an advantage, the leaders of all your three divisions will fall into the hands of the enemy.

8. The stronger men will be in front, the jaded ones will fall behind, and on this plan only one-tenth of your army will reach its destination.

9. If you march fifty LI in order to outmaneuver the enemy, you will lose the leader of your first division, and only half your force will reach the goal.

10. If you march thirty LI with the same object, two-thirds of your army will arrive.

11. We may take it then that an army without its baggage-train is lost; without provisions it is lost; without bases of supply it is lost.

12. We cannot enter into alliances until we are acquainted with the designs of our neighbors.

13. We are not fit to lead an army on the march unless we are familiar with the face of the country - its mountains and forests, its pitfalls and precipices, its marshes and swamps.

14. We shall be unable to turn natural advantage to account unless we make use of local guides.

15. In war, practice dissimulation, and you will succeed.

16. Whether to concentrate or to divide your troops, must be decided by circumstances.

17. Let your rapidity be that of the wind, your compactness that of the forest.

18. In raiding and plundering be like fire, in immovability like a mountain.

19. Let your plans be dark and impenetrable as night, and when you move, fall like a thunderbolt.

20. When you plunder a countryside, let the spoil be divided amongst your men; when you capture new territory, cut it up into allotments for the benefit of the soldiery.

21. Ponder and deliberate before you make a move.

22. He will conquer who has learnt the artifice of deviation. Such is the art of maneuvering.

23. The Book of Army Management says: On the field of battle, the spoken word does not carry far enough: hence the institution of gongs and drums. Nor can ordinary objects be seen clearly enough: hence the institution of banners and flags.

24. Gongs and drums, banners and flags, are means whereby the ears and eyes of the host may be focused on one particular point.

25. The host thus forming a single united body, is it impossible either for the brave to advance alone, or for the cowardly to retreat alone. This is the art of handling large masses of men.

26. In night-fighting, then, make much use of signal-fires and drums, and in fighting by day, of flags and banners, as a means of influencing the ears and eyes of your army.

27. A whole army may be robbed of its spirit; a commander-in-chief may be robbed of his presence of mind.

28. Now a soldier's spirit is keenest in the morning; by noonday it has begun to flag; and in the evening, his mind is bent only on returning to camp.

29. A clever general, therefore, avoids an army when its spirit is keen, but attacks it when it is sluggish and inclined to return. This is the art of studying moods.

30. Disciplined and calm, to await the appearance of disorder and hubbub amongst the enemy:- this is the art of retaining self-possession.

31. To be near the goal while the enemy is still far from it, to wait at ease while the enemy is toiling and struggling, to be

well-fed while the enemy is famished:- this is the art of husbanding one's strength.

32. To refrain from intercepting an enemy whose banners are in perfect order, to refrain from attacking an army drawn up in calm and confident array:- this is the art of studying circumstances.

33. It is a military axiom not to advance uphill against the enemy, nor to oppose him when he comes downhill.

34. Do not pursue an enemy who simulates flight; do not attack soldiers whose temper is keen.

35. Do not swallow bait offered by the enemy. Do not interfere with an army that is returning home.

36. When you surround an army, leave an outlet free. Do not press a desperate foe too hard.

37. Such is the art of warfare.

8
VARIATION IN TACTICS

1. Sun Tzu said: In war, the general receives his commands from the sovereign, collects his army and concentrates his forces

2. When in difficult country, do not encamp. In country where high roads intersect, join hands with your allies. Do not linger in dangerously isolated positions. In hemmed-in situations, you must resort to stratagem. In desperate position, you must fight.

3. There are roads which must not be followed, armies which must be not attacked, towns which must not be besieged, positions which must not be contested, commands of the sovereign which must not be obeyed.

4. The general who thoroughly understands the advantages that accompany variation of tactics knows how to handle his troops.

5. The general who does not understand these, may be well acquainted with the configuration of the country, yet he will not be able to turn his knowledge to practical account.

6. So, the student of war who is unversed in the art of war of varying his plans, even though he be acquainted with the Five Advantages, will fail to make the best use of his men.

7. Hence in the wise leader's plans, considerations of advantage and of disadvantage will be blended together.

8. If our expectation of advantage be tempered in this way, we may succeed in accomplishing the essential part of our schemes.

9. If, on the other hand, in the midst of difficulties we are always ready to seize an advantage, we may extricate ourselves from misfortune.

10. Reduce the hostile chiefs by inflicting damage on them; and make trouble for them, and keep them constantly engaged; hold out specious allurements, and make them rush to any given point.

11. The art of war teaches us to rely not on the likelihood of the enemy's not coming, but on our own readiness to receive him; not on the chance of his not attacking, but rather on the fact that we have made our position unassailable.

12. There are five dangerous faults which may affect a general: (1) Recklessness, which leads to destruction; (2) cowardice, which leads to capture; (3) a hasty temper, which

can be provoked by insults; (4) a delicacy of honor which is sensitive to shame; (5) over-solicitude for his men, which exposes him to worry and trouble.

13. These are the five besetting sins of a general, ruinous to the conduct of war.

14. When an army is overthrown and its leader slain, the cause will surely be found among these five dangerous faults. Let them be a subject of meditation.

9
THE ARMY ON THE MARCH

1. Sun Tzu said: We come now to the question of encamping the army, and observing signs of the enemy. Pass quickly over mountains, and keep in the neighborhood of valleys.

2. Camp in high places, facing the sun. Do not climb heights in order to fight. So much for mountain warfare.

3. After crossing a river, you should get far away from it.

4. When an invading force crosses a river in its onward march, do not advance to meet it in mid-stream. It will be best to let half the army get across, and then deliver your attack.

5. If you are anxious to fight, you should not go to meet the invader near a river which he has to cross.

6. Moor your craft higher up than the enemy, and facing the sun. Do not move up-stream to meet the enemy. So much for river warfare.

7. In crossing salt-marshes, your sole concern should be to get over them quickly, without any delay.

8. If forced to fight in a salt-marsh, you should have water and grass near you, and get your back to a clump of trees. So much for operations in salt-marshes.

9. In dry, level country, take up an easily accessible position with rising ground to your right and on your rear, so that the danger may be in front, and safety lie behind. So much for campaigning in flat country.

10. These are the four useful branches of military knowledge which enabled the Yellow Emperor to vanquish four several sovereigns.

11. All armies prefer high ground to low and sunny places to dark.

12. If you are careful of your men, and camp on hard ground, the army will be free from disease of every kind, and this will spell victory.

13. When you come to a hill or a bank, occupy the sunny side, with the slope on your right rear. Thus you will at once act for the benefit of your soldiers and utilize the natural advantages of the ground.

14. When, in consequence of heavy rains up-country, a river which you wish to ford is swollen and flecked with foam, you must wait until it subsides.

15. Country in which there are precipitous cliffs with torrents running between, deep natural hollows, confined places, tangled thickets, quagmires and crevasses, should be left with all possible speed and not approached.

16. While we keep away from such places, we should get the enemy to approach them; while we face them, we should let the enemy have them on his rear.

17. If in the neighborhood of your camp there should be any hilly country, ponds surrounded by aquatic grass, hollow basins filled with reeds, or woods with thick undergrowth, they must be carefully routed out and searched; for these are places where men in ambush or insidious spies are likely to be lurking.

18. When the enemy is close at hand and remains quiet, he is relying on the natural strength of his position.

19. When he keeps aloof and tries to provoke a battle, he is anxious for the other side to advance.

20. If his place of encampment is easy of access, he is tendering a bait.

21. Movement amongst the trees of a forest shows that the enemy is advancing. The appearance of a number of screens in the midst of thick grass means that the enemy wants to make us suspicious.

22. The rising of birds in their flight is the sign of an ambuscade. Startled beasts indicate that a sudden attack is coming.

23. When there is dust rising in a high column, it is the sign of chariots advancing; when the dust is low, but spread over a wide area, it betokens the approach of infantry. When it branches out in different directions, it shows that parties have been sent to collect firewood. A few clouds of dust moving to and fro signify that the army is encamping.

24. Humble words and increased preparations are signs that the enemy is about to advance. Violent language and driving forward as if to the attack are signs that he will retreat.

25. When the light chariots come out first and take up a position on the wings, it is a sign that the enemy is forming for battle.

26. Peace proposals unaccompanied by a sworn covenant indicate a plot.

27. When there is much running about and the soldiers fall into rank, it means that the critical moment has come.

28. When some are seen advancing and some retreating, it is a lure.

29. When the soldiers stand leaning on their spears, they are faint from want of food.

30. If those who are sent to draw water begin by drinking themselves, the army is suffering from thirst.

31. If the enemy sees an advantage to be gained and makes no effort to secure it, the soldiers are exhausted.

32. If birds gather on any spot, it is unoccupied. Clamor by night betokens nervousness.

33. If there is disturbance in the camp, the general's authority is weak. If the banners and flags are shifted about, sedition is afoot. If the officers are angry, it means that the men are weary.

34. When an army feeds its horses with grain and kills its cattle for food, and when the men do not hang their cooking-pots over the camp-fires, showing that they will not return to their tents, you may know that they are determined to fight to the death.

35. The sight of men whispering together in small knots or speaking in subdued tones points to disaffection amongst the rank and file.

36. Too frequent rewards signify that the enemy is at the end of his resources; too many punishments betray a condition of dire distress.

37. To begin by bluster, but afterwards to take fright at the enemy's numbers, shows a supreme lack of intelligence.

38. When envoys are sent with compliments in their mouths, it is a sign that the enemy wishes for a truce.

39. If the enemy's troops march up angrily and remain facing ours for a long time without either joining battle or taking themselves off again, the situation is one that demands great vigilance and circumspection.

40. If our troops are no more in number than the enemy, that is amply sufficient; it only means that no direct attack can be made. What we can do is simply to concentrate all our available strength, keep a close watch on the enemy, and obtain reinforcements.

41. He who exercises no forethought but makes light of his opponents is sure to be captured by them.

42. If soldiers are punished before they have grown attached to you, they will not prove submissive; and, unless submissive, then will be practically useless. If, when the soldiers have become attached to you, punishments are not enforced, they will still be unless.

43. Therefore soldiers must be treated in the first instance with humanity, but kept under control by means of iron discipline. This is a certain road to victory.

44. If in training soldiers commands are habitually enforced, the army will be well-disciplined; if not, its discipline will be bad.

45. If a general shows confidence in his men but always insists on his orders being obeyed, the gain will be mutual.

10
Terrain

1. Sun Tzu said: We may distinguish six kinds of terrain, to wit: (1) Accessible ground; (2) entangling ground; (3) temporizing ground; (4) narrow passes; (5) precipitous heights; (6) positions at a great distance from the enemy.

2. Ground which can be freely traversed by both sides is called accessible.

3. With regard to ground of this nature, be before the enemy in occupying the raised and sunny spots, and carefully guard your line of supplies. Then you will be able to fight with advantage.

4. Ground which can be abandoned but is hard to re-occupy is called entangling.

5. From a position of this sort, if the enemy is unprepared, you may sally forth and defeat him. But if the enemy is prepared for your coming, and you fail to defeat him, then, return being impossible, disaster will ensue.

6. When the position is such that neither side will gain by making the first move, it is called temporizing ground.

7. In a position of this sort, even though the enemy should offer us an attractive bait, it will be advisable not to stir forth, but rather to retreat, thus enticing the enemy in his turn; then, when part of his army has come out, we may deliver our attack with advantage.

8. With regard to narrow passes, if you can occupy them first, let them be strongly garrisoned and await the advent of the enemy.

9. Should the army forestall you in occupying a pass, do not go after him if the pass is fully garrisoned, but only if it is weakly garrisoned.

10. With regard to precipitous heights, if you are beforehand with your adversary, you should occupy the raised and sunny spots, and there wait for him to come up.

11. If the enemy has occupied them before you, do not follow him, but retreat and try to entice him away.

12. If you are situated at a great distance from the enemy, and the strength of the two armies is equal, it is not easy to provoke a battle, and fighting will be to your disadvantage.

13. These six are the principles connected with Earth. The general who has attained a responsible post must be careful to study them.

14. Now an army is exposed to six several calamities, not arising from natural causes, but from faults for which the general is responsible. These are: (1) Flight; (2) Insubordination; (3) Collapse; (4) Ruin; (5) Disorganization; (6) Rout.

15. Other conditions being equal, if one force is hurled against another ten times its size, the result will be the Flight of the former.

16. When the common soldiers are too strong and their officers too weak, the result is Insubordination. When the officers are too strong and the common soldiers too weak, the result is Collapse.

17. When the higher officers are angry and insubordinate, and on meeting the enemy give battle on their own account from a feeling of resentment, before the commander-in-chief can tell whether or no he is in a position to fight, the result is ruin.

18. When the general is weak and without authority; when his orders are not clear and distinct; when there are no fixed duties assigned to officers and men, and the ranks are formed in a slovenly haphazard manner, the result is utter Disorganization.

19. When a general, unable to estimate the enemy's strength, allows an inferior force to engage a larger one, or hurls a weak detachment against a powerful one, and neglects to

place picked soldiers in the front rank, the result must be Rout.

20. These are six ways of courting defeat, which must be carefully noted by the general who has attained a responsible post.

21. The natural formation of the country is the soldier's best ally; but a power of estimating the adversary, of controlling the forces of victory, and of shrewdly calculating difficulties, dangers and distances, constitutes the test of a great general.

22. He who knows these things, and in fighting puts his knowledge into practice, will win his battles. He who knows them not, nor practices them, will surely be defeated.

23. If fighting is sure to result in victory, then you must fight, even though the ruler forbid it; if fighting will not result in victory, then you must not fight even at the ruler's bidding.

24. The general who advances without coveting fame and retreats without fearing disgrace, whose only thought is to protect his country and do good service for his sovereign, is the jewel of the kingdom.

25. Regard your soldiers as your children, and they will follow you into the deepest valleys; look upon them as your own beloved sons, and they will stand by you even unto death.

26. If, however, you are indulgent, but unable to make your authority felt; kind-hearted, but unable to enforce your commands; and incapable, moreover, of quelling disorder: then your soldiers must be likened to spoilt children; they are useless for any practical purpose.

27. If we know that our own men are in a condition to attack, but are unaware that the enemy is not open to attack, we have gone only halfway towards victory.

28. If we know that the enemy is open to attack, but are unaware that our own men are not in a condition to attack, we have gone only halfway towards victory.

29. If we know that the enemy is open to attack, and also know that our men are in a condition to attack, but are unaware that the nature of the ground makes fighting impracticable, we have still gone only halfway towards victory.

30. Hence the experienced soldier, once in motion, is never bewildered; once he has broken camp, he is never at a loss.

31. Hence the saying: If you know the enemy and know yourself, your victory will not stand in doubt; if you know Heaven and know Earth, you may make your victory complete.

11
THE NINE SITUATIONS

1. Sun Tzu said: The art of war recognizes nine varieties of ground: (1) Dispersive ground; (2) facile ground; (3) contentious ground; (4) open ground; (5) ground of intersecting highways; (6) serious ground; (7) difficult ground; (8) hemmed-in ground; (9) desperate ground.

2. When a chieftain is fighting in his own territory, it is dispersive ground.

3. When he has penetrated into hostile territory, but to no great distance, it is facile ground.

4. Ground the possession of which imports great advantage to either side, is contentious ground.

5. Ground on which each side has liberty of movement is open ground.

6. Ground which forms the key to three contiguous states, so that he who occupies it first has most of the Empire at his command, is a ground of intersecting highways.

7. When an army has penetrated into the heart of a hostile country, leaving a number of fortified cities in its rear, it is serious ground.

8. Mountain forests, rugged steeps, marshes and fens--all country that is hard to traverse: this is difficult ground.

9. Ground which is reached through narrow gorges, and from which we can only retire by tortuous paths, so that a small number of the enemy would suffice to crush a large body of our men: this is hemmed in ground.

10. Ground on which we can only be saved from destruction by fighting without delay, is desperate ground.

11. On dispersive ground, therefore, fight not. On facile ground, halt not. On contentious ground, attack not.

12. On open ground, do not try to block the enemy's way. On the ground of intersecting highways, join hands with your allies.

13. On serious ground, gather in plunder. In difficult ground, keep steadily on the march.

14. On hemmed-in ground, resort to stratagem. On desperate ground, fight.

15. Those who were called skillful leaders of old knew how to drive a wedge between the enemy's front and rear; to prevent co-operation between his large and small divisions; to hinder the good troops from rescuing the bad, the officers from rallying their men.

16. When the enemy's men were united, they managed to keep them in disorder.

17. When it was to their advantage, they made a forward move; when otherwise, they stopped still.

18. If asked how to cope with a great host of the enemy in orderly array and on the point of marching to the attack, I should say: "Begin by seizing something which your opponent holds dear; then he will be amenable to your will."

19. Rapidity is the essence of war: take advantage of the enemy's unreadiness, make your way by unexpected routes, and attack unguarded spots.

20. The following are the principles to be observed by an invading force: The further you penetrate into a country, the greater will be the solidarity of your troops, and thus the defenders will not prevail against you.

21. Make forays in fertile country in order to supply your army with food.

22. Carefully study the well-being of your men, and do not overtax them. Concentrate your energy and hoard your

strength. Keep your army continually on the move, and devise unfathomable plans.

23. Throw your soldiers into positions whence there is no escape, and they will prefer death to flight. If they will face death, there is nothing they may not achieve. Officers and men alike will put forth their uttermost strength.

24. Soldiers when in desperate straits lose the sense of fear. If there is no place of refuge, they will stand firm. If they are in hostile country, they will show a stubborn front. If there is no help for it, they will fight hard.

25. Thus, without waiting to be marshaled, the soldiers will be constantly on the qui vive; without waiting to be asked, they will do your will; without restrictions, they will be faithful; without giving orders, they can be trusted.

26. Prohibit the taking of omens, and do away with superstitious doubts. Then, until death itself comes, no calamity need be feared.

27. If our soldiers are not overburdened with money, it is not because they have a distaste for riches; if their lives are not unduly long, it is not because they are disinclined to longevity.

28. On the day they are ordered out to battle, your soldiers may weep, those sitting up bedewing their garments, and those lying down letting the tears run down their cheeks. But let them once be brought to bay, and they will display the courage of a Chu or a Kuei.

29. The skillful tactician may be likened to the Shuai-Jan. Now the Shuai-Jan is a snake that is found in the Ch`ang mountains. Strike at its head, and you will be attacked by its tail; strike at its tail, and you will be attacked by its head; strike at its middle, and you will be attacked by head and tail both.

30. Asked if an army can be made to imitate the Shuai-Jan, I should answer, Yes. For the men of Wu and the men of Yueh are enemies; yet if they are crossing a river in the same boat and are caught by a storm, they will come to each other's assistance just as the left hand helps the right.

31. Hence it is not enough to put one's trust in the tethering of horses, and the burying of chariot wheels in the ground

32. The principle on which to manage an army is to set up one standard of courage which all must reach.

33. How to make the best of both strong and weak--that is a question involving the proper use of ground.

34. Thus the skillful general conducts his army just as though he were leading a single man, willy-nilly, by the hand.

35. It is the business of a general to be quiet and thus ensure secrecy; upright and just, and thus maintain order.

36. He must be able to mystify his officers and men by false reports and appearances, and thus keep them in total ignorance.

37. By altering his arrangements and changing his plans, he keeps the enemy without definite knowledge. By shifting his camp and taking circuitous routes, he prevents the enemy from anticipating his purpose.

38. At the critical moment, the leader of an army acts like one who has climbed up a height and then kicks away the ladder behind him. He carries his men deep into hostile territory before he shows his hand.

39. He burns his boats and breaks his cooking-pots; like a shepherd driving a flock of sheep, he drives his men this way and that, and nothing knows whither he is going.

40. To muster his host and bring it into danger:- this may be termed the business of the general.

41. The different measures suited to the nine varieties of ground; the expediency of aggressive or defensive tactics; and the fundamental laws of human nature: these are things that must most certainly be studied.

42. When invading hostile territory, the general principle is, that penetrating deeply brings cohesion; penetrating but a short way means dispersion.

43. When you leave your own country behind, and take your army across neighborhood territory, you find yourself on

critical ground. When there are means of communication on all four sides, the ground is one of intersecting highways.

44. When you penetrate deeply into a country, it is serious ground. When you penetrate but a little way, it is facile ground.

45. When you have the enemy's strongholds on your rear, and narrow passes in front, it is hemmed-in ground. When there is no place of refuge at all, it is desperate ground.

46. Therefore, on dispersive ground, I would inspire my men with unity of purpose. On facile ground, I would see that there is close connection between all parts of my army.

47. On contentious ground, I would hurry up my rear.

48. On open ground, I would keep a vigilant eye on my defenses. On ground of intersecting highways, I would consolidate my alliances.

49. On serious ground, I would try to ensure a continuous stream of supplies. On difficult ground, I would keep pushing on along the road.

50. On hemmed-in ground, I would block any way of retreat. On desperate ground, I would proclaim to my soldiers the hopelessness of saving their lives.

51. For it is the soldier's disposition to offer an obstinate resistance when surrounded, to fight hard when he cannot

help himself, and to obey promptly when he has fallen into danger.

52. We cannot enter into alliance with neighboring princes until we are acquainted with their designs. We are not fit to lead an army on the march unless we are familiar with the face of the country - its mountains and forests, its pitfalls and precipices, its marshes and swamps. We shall be unable to turn natural advantages to account unless we make use of local guides.

53. To be ignored of any one of the following four or five principles does not befit a warlike prince.

54. When a warlike prince attacks a powerful state, his generalship shows itself in preventing the concentration of the enemy's forces. He overawes his opponents, and their allies are prevented from joining against him.

55. Hence he does not strive to ally himself with all and sundry, nor does he foster the power of other states. He carries out his own secret designs, keeping his antagonists in awe. Thus he is able to capture their cities and overthrow their kingdoms.

56. Bestow rewards without regard to rule, issue orders without regard to previous arrangements; and you will be able to handle a whole army as though you had to do with but a single man.

57. Confront your soldiers with the deed itself; never let them know your design. When the outlook is bright, bring it

before their eyes; but tell them nothing when the situation is gloomy.

58. Place your army in deadly peril, and it will survive; plunge it into desperate straits, and it will come off in safety.

59. For it is precisely when a force has fallen into harm's way that is capable of striking a blow for victory.

60. Success in warfare is gained by carefully accommodating ourselves to the enemy's purpose.

61. By persistently hanging on the enemy's flank, we shall succeed in the long run in killing the commander-in-chief.

62. This is called ability to accomplish a thing by sheer cunning.

63. On the day that you take up your command, block the frontier passes, destroy the official tallies, and stop the passage of all emissaries.

64. Be stern in the council-chamber, so that you may control the situation.

65. If the enemy leaves a door open, you must rush in.

66. Forestall your opponent by seizing what he holds dear, and subtly contrive to time his arrival on the ground.

67. Walk in the path defined by rule, and accommodate yourself to the enemy until you can fight a decisive battle.

68. At first, then, exhibit the coyness of a maiden, until the enemy gives you an opening; afterwards emulate the rapidity of a running hare, and it will be too late for the enemy to oppose you.

12
THE ATTACK BY FIRE

1. Sun Tzu said: There are five ways of attacking with fire. The first is to burn soldiers in their camp; the second is to burn stores; the third is to burn baggage trains; the fourth is to burn arsenals and magazines; the fifth is to hurl dropping fire amongst the enemy.

2. In order to carry out an attack, we must have means available. The material for raising fire should always be kept in readiness.

3. There is a proper season for making attacks with fire, and special days for starting a conflagration.

4. The proper season is when the weather is very dry; the special days are those when the moon is in the constellations of the Sieve, the Wall, the Wing or the Cross-bar; for these four are all days of rising wind.

5. In attacking with fire, one should be prepared to meet five possible developments:

6. (1) When fire breaks out inside to enemy's camp, respond at once with an attack from without.

7. (2) If there is an outbreak of fire, but the enemy's soldiers remain quiet, bide your time and do not attack.

8. (3) When the force of the flames have reached its height, follow it up with an attack, if that is practicable; if not, stay where you are.

9. (4) If it is possible to make an assault with fire from without, do not wait for it to break out within, but deliver your attack at a favorable moment.

10. (5) When you start a fire, be to windward of it. Do not attack from the leeward.

11. A wind that rises in the daytime lasts long, but a night breeze soon falls.

12. In every army, the five developments connected with fire must be known, the movements of the stars calculated, and a watch kept for the proper days.

13. Hence those who use fire as an aid to the attack show intelligence; those who use water as an aid to the attack gain an accession of strength.

14. By means of water, an enemy may be intercepted, but not robbed of all his belongings.

15. Unhappy is the fate of one who tries to win his battles and succeed in his attacks without cultivating the spirit of enterprise; for the result is waste of time and general stagnation.

16. Hence the saying: The enlightened ruler lays his plans well ahead; the good general cultivates his resources.

17. Move not unless you see an advantage; use not your troops unless there is something to be gained; fight not unless the position is critical.

18. No ruler should put troops into the field merely to gratify his own spleen; no general should fight a battle simply out of pique.

19. If it is to your advantage, make a forward move; if not, stay where you are.

20. Anger may in time change to gladness; vexation may be succeeded by content.

21. But a kingdom that has once been destroyed can never come again into being; nor can the dead ever be brought back to life.

22. Hence the enlightened ruler is heedful, and the good general full of caution. This is the way to keep a country at peace and an army intact.

13
THE USE OF SPIES

1. Sun Tzu said: Raising a host of a hundred thousand men and marching them great distances entails heavy loss on the people and a drain on the resources of the State. The daily expenditure will amount to a thousand ounces of silver. There will be commotion at home and abroad, and men will drop down exhausted on the highways. As many as seven hundred thousand families will be impeded in their labor.

2. Hostile armies may face each other for years, striving for the victory which is decided in a single day. This being so, to remain in ignorance of the enemy's condition simply because one grudges the outlay of a hundred ounces of silver in honors and emoluments, is the height of inhumanity.

3. One who acts thus is no leader of men, no present help to his sovereign, no master of victory.

4. Thus, what enables the wise sovereign and the good general to strike and conquer, and achieve things beyond the reach of ordinary men, is foreknowledge.

5. Now this foreknowledge cannot be elicited from spirits; it cannot be obtained inductively from experience, nor by any deductive calculation.

6. Knowledge of the enemy's dispositions can only be obtained from other men.

7. Hence the use of spies, of whom there are five classes: (1) Local spies; (2) inward spies; (3) converted spies; (4) doomed spies; (5) surviving spies.

8. When these five kinds of spy are all at work, none can discover the secret system. This is called "divine manipulation of the threads." It is the sovereign's most precious faculty.

9. Having local spies means employing the services of the inhabitants of a district.

10. Having inward spies, making use of officials of the enemy.

11. Having converted spies, getting hold of the enemy's spies and using them for our own purposes.

12. Having doomed spies, doing certain things openly for purposes of deception, and allowing our spies to know of them and report them to the enemy.

13. Surviving spies, finally, are those who bring back news from the enemy's camp.

14. Hence it is that which none in the whole army are more intimate relations to be maintained than with spies. None should be more liberally rewarded. In no other business should greater secrecy be preserved.

15. Spies cannot be usefully employed without a certain intuitive sagacity.

16. They cannot be properly managed without benevolence and straightforwardness.

17. Without subtle ingenuity of mind, one cannot make certain of the truth of their reports.

18. Be subtle! be subtle! and use your spies for every kind of business.

19. If a secret piece of news is divulged by a spy before the time is ripe, he must be put to death together with the man to whom the secret was told.

20. Whether the object be to crush an army, to storm a city, or to assassinate an individual, it is always necessary to begin by finding out the names of the attendants, the aides-de-camp, and door-keepers and sentries of the general in command. Our spies must be commissioned to ascertain these.

21. The enemy's spies who have come to spy on us must be sought out, tempted with bribes, led away and comfortably housed. Thus they will become converted spies and available for our service.

22. It is through the information brought by the converted spy that we are able to acquire and employ local and inward spies.

23. It is owing to his information, again, that we can cause the doomed spy to carry false tidings to the enemy.

24. Lastly, it is by his information that the surviving spy can be used on appointed occasions.

25. The end and aim of spying in all its five varieties is knowledge of the enemy; and this knowledge can only be derived, in the first instance, from the converted spy. Hence it is essential that the converted spy be treated with the utmost liberality.

26. Of old, the rise of the Yin dynasty was due to I Chih who had served under the Hsia. Likewise, the rise of the Chou dynasty was due to Lu Ya who had served under the Yin.

27. Hence it is only the enlightened ruler and the wise general who will use the highest intelligence of the army for purposes of spying and thereby they achieve great results. Spies are a most important element in war, because on them depends an army's ability to move.

www.ingramcontent.com/pod-product-compliance
Lightning Source LLC
Chambersburg PA
CBHW050508210326
41521CB00011B/2367